NUMERACY AS SOCIAL PRACTICE

Learning takes place both inside and outside of the classroom, embedded in local practices, traditions and interactions. But whereas the importance of social practice is increasingly recognised in literacy education, *Numeracy as Social Practice: Global and Local Perspectives* is the first book to fully explore these principles in the context of numeracy. The book brings together a wide range of accounts and studies from around the world to build a picture of the challenges and benefits of seeing numeracy as social practice – that is, as mathematical activities embedded in the social, cultural, historical and political contexts in which these activities take place.

Drawing on workplace, community and classroom contexts, *Numeracy as Social Practice* shows how everyday numeracy practices can be used in formal and non-formal maths teaching and how, in turn, classroom teaching can help to validate and strengthen local numeracy practices. At a time when an increasingly transnational approach is taken to education policy making, this book will appeal to development practitioners and researchers, and adult education, mathematics and numeracy teachers, researchers and policy makers around the world.

Keiko Yasukawa is an adult numeracy and literacy researcher and teacher educator at the University of Technology Sydney in Australia.

Alan Rogers is an adult educator and Visiting Professor at the universities of East Anglia and Nottingham, UK.

Kara Jackson is an Associate Professor of Mathematics Education at the University of Washington, Seattle, USA.

Brian V. Street was an anthropologist, formerly Professor of Language in Education at King's College, London, UK, and Visiting Professor of Education at the University of Pennsylvania, USA.

Rethinking Development

Rethinking Development offers accessible and thought-provoking overviews of contemporary topics in international development and aid. Providing original empirical and analytical insights, the books in this series push thinking in new directions by challenging current conceptualizations and developing new ones.

This is a dynamic and inspiring series for all those engaged with today's debates surrounding development issues, whether they be students, scholars, policy makers and practitioners internationally. These interdisciplinary books provide an invaluable resource for discussion in advanced undergraduate and postgraduate courses in development studies as well as in anthropology, economics, politics, geography, media studies and sociology.

Celebrity Humanitarianism and North–South Relations
Politics, Place and Power
Edited by Lisa-Ann Richey

Education, Learning and the Transformation of Development
Edited by Amy Skinner, Matt Baillie Smith, Eleanor Brown and Tobias Troll

Learning and Volunteering Abroad for Development
Unpacking Host Organisation and Volunteer Rationales
Rebecca Tiessen

Communicating Development with Communities
Linje Manyozo

Learning and Volunteering Abroad for Development
Emerging Economies and Development
Jan Nederveen Pieterse

Disability and International Development
A Guide for Students and Practitioners
David Cobley

Communication in International Development
Doing Good or Looking Good?
Edited by Florencia Enghel and Jessica Noske-Turner

Epistemic Freedom in Africa
Deprovincialization and Decolonization
Sabelo J. Ndlovu-Gatsheni

NUMERACY AS SOCIAL PRACTICE

Global and Local Perspectives

Edited by Keiko Yasukawa, Alan Rogers,
Kara Jackson and Brian V. Street

Routledge
Taylor & Francis Group

LONDON AND NEW YORK

First published 2018
by Routledge
2 Park Square, Milton Park, Abingdon, Oxon OX14 4RN

and by Routledge
711 Third Avenue, New York, NY 10017

Routledge is an imprint of the Taylor & Francis Group, an informa business

© 2018 selection and editorial matter, Keiko Yasukawa, Alan
Rogers, Kara Jackson and Brian V. Street; individual chapters, the
contributors

British Library Cataloguing-in-Publication Data
A catalogue record for this book is available from the British Library

Library of Congress Cataloging-in-Publication Data
Names: Yasukawa, Keiko, editor.
Title: Numeracy as social practice : global and local perspectives /
edited by Keiko Yasukawa, Alan Rogers, Kara Jackson and Brian V.
Street.
Description: Abingdon, Oxon ; New York, NY : Routledge, 2018. |
Includes bibliographical references.
Identifiers: LCCN 2017058016| ISBN 9781138284449 (hardback)
| ISBN 9781138284456 (pbk.) | ISBN 9781315269474 (ebook)
Subjects: LCSH: Numeracy – Study and teaching – Social aspects –
Case studies. | Numeracy – Social aspects – Case studies.
Classification: LCC QA141 .N844 2018 | DDC 513.071–dc23
LC record available at https://lccn.loc.gov/2017058016

ISBN: 978-1-138-28444-9 (hbk)
ISBN: 978-1-138-28445-6 (pbk)
ISBN: 978-1-315-26947-4 (ebk)

Typeset in Bembo
by HWA Text and Data Management, London

Professor Brian V. Street, one of the co-editors of this volume, died on 21 June 2017 during the final stages of the preparation of this book.

The remaining co-editors wish to dedicate this volume to the memory of

BRIAN V. STREET (1943–2017)

inspirational scholar, teacher and friend

'The volume brings numeracy as social practice to life in ethnographic case studies of everyday numeracy practices and mathematics education in diverse international settings. Different theoretical perspectives are woven across the chapters, including new literacy studies, cultural-historical activity theory, critical theory and ethnomathematics. The editors masterfully craft all this into a coherent volume useful to researchers and mathematics educators around the world.'

Stephen Reder, Portland State University, USA

'This lucid account of numeracy as social practice is much needed and long overdue. It features an admirably broad scope and diversity, with chapters from every continent and from various educational, work, and other everyday settings. I especially like the discussions on relations between everyday numeracy practices and more formal mathematics, the invisibility of numeracy practices, and the importance of attending to power relations.'

Jeff Evans, Reader Emeritus, Middlesex University, London, UK

'This pioneering publication is timely in that the importance of learning as a social practice, especially numeracy as a social practice, is increasingly being understood and recognized by those concerned with quality education, formal/non-formal/lifelong/life-wide education. Rooted in a number of relevant research studies from around the world, and informed by theoretical influences highlighting the "social practices perspective on numeracy", the authors engage the educators and researchers in broadening their vision of numeracy. This is a valuable resource book for researchers, activists and educators.'

L. S. Saraswathi, Social Activist, Trainer and Researcher in Development, Chennai, India

CONTENTS

ILLUSTRATIONS

Figures

Table

CONTRIBUTORS

Wilfredo Vidal Alangui is a professor in the Department of Mathematics and Computer Science at the University of the Philippines Baguio, Baguio City, Philippines. His research interest is on the interplay of mathematics and culture, with a focus on indigenous peoples' education.

Jehad Alshwaikh is an assistant professor in mathematics education and teacher educator at Birzeit University in Palestine. His interest is in representation and communication in mathematical discourse.

Richard Barwell is Professor of Mathematics Education at the University of Ottawa, Canada. His research concerns questions relating to the role of language and social interaction in mathematics classrooms and in the discourse of mathematics education. His research is informed by ideas in discursive psychology, Bakhtinian sociolinguistics and linguistic ethnography.

Elisabet Bellander is a teacher in mathematics and science, and works as an upper secondary school teacher in mathematics at Universitetsholmens Gymnasium in Malmö, Sweden. Together with Michael Blaesild, Elisabet runs a project on mathematics in relation to other teaching contents, such as vocational curricula. They are facilitators for collaborations between mathematics teachers and vocational teachers.

Michael Blaesild is qualified as a construction worker and vocational teacher. He works as an upper secondary school teacher in construction work at Byggymnasiet in Malmö, Sweden. Together with Elisabet Bellander, Michael runs a project on mathematics in relation to other teaching contents, such as vocational curricula. They are facilitators for collaborations between mathematics teachers and vocational teachers.

Lisa Björklund Boistrup is a senior lecturer at Stockholm University, Sweden. One of her research areas is to understand mathematics-containing activities outside school such as in workplaces. Her most recent works include a book chapter 'Assessment in mathematics education: A gatekeeping dispositive' (in Straehler et al. (eds), *The Disorder of Mathematics Education*, 2017).

Diana Coben is Director of the National Centre of Literacy and Numeracy for Adults at The University of Waikato, New Zealand and Emeritus Professor of Adult Numeracy at King's College London. Diana was Founding Chair and is now an honorary trustee of Adults Learning Mathematics – A Research Forum (ALM), an international research forum bringing together researchers and practitioners in adult mathematics/numeracy teaching and learning in order to promote the learning of mathematics by adults.

Kara Jackson is an associate professor of mathematics education at the University of Washington, Seattle, USA. Her research focuses on understanding how to improve mathematics teaching and learning – especially in the middle grades – to support youth from historically underserved communities to participate substantially in and identify with academically rigorous mathematics.

Judy Kalman is professor in the Departamento de Investigaciones Educativas in the Centro de Investigacion y Estudios Avanzados in Mexico City, Mexico. Her work focuses on social processes underlying literacy and numeracy, particularly among marginalised adults and youths. She has published extensively on literacy and numeracy in Latin America and is a winner of the UNESCO Prize for Literacy.

Phil Kane is a professional teaching fellow at the University of Auckland, New Zealand. Research interests include how people use estimation and spatial thinking as elements of workplace numeracy, and more recently how pre-degree or bridging learners contend with having to repeat a mathematics entry course.

Herbert Khuzwayo is an associate professor in the School of Science and Mathematics Education at the University of the Western Cape, South Africa. His research interests include the following areas: history of mathematics teaching and learning, mathematics teacher development, and social and political perspectives in mathematics education.

Gelsa Knijnik is a professor at the Graduate Program on Education of Unisinos, in Brazil. For decades she has been working with peasants of the Brazilian

Landless Movement. Her publications are known both in Latin America and more generally.

Rebecca Nthogo Lekoko is a professor of adult education in the Department of Adult Education, University of Botswana specialising in community development and curriculum issues. Her professional development in these areas can be explained as weaving together experiences in diverse areas of indigenous knowledge, the marginalised, out-of-school education, non-formal education and open and distance learning.

Obusitswe Pitso is a mathematics education lecturer in the Department of Mathematics and Science Education, University of Botswana, but prior to that he was a curriculum development and evaluation specialist in the Ministry of Education, Botswana. His main research areas are in ethnomathematics, mathematics curriculum development, planning, designing and implementation.

Anita Rampal is professor and former Dean, Faculty of Education, Delhi University. Her special interests include policy analysis, curriculum studies, science-technology-society (STS) education and critical mathematics education.

Alan Rogers is an adult educator with research and training interests in adult learning and teaching. He has worked widely as a practitioner in the training of teachers in adult literacy and basic education, especially in countries of Africa and Asia. He recently completed a review of the folk development colleges in Tanzania (with the assistance of Sida) and is currently working on projects in Ethiopia and Afghanistan. A visiting professor at the universities of East Anglia and Nottingham, UK, he has written extensively on adult literacy and basic education, and practitioner training.

Mariko Shiohata works for Save the Children Japan as Director of International Programmes. Between 2012 and 2015, she ran a three-year basic education project with the Ministry of Education of Nepal.

Diana Solares has done extensive work on numeracy among itinerate/migrant farm workers in Mexico. She is currently teaching at the Universidad Autónoma de Queretaro, Mexico.

Brian V. Street did anthropological field work in Iran during the 1970s, from which he developed theoretical approaches to literacy in cross-cultural perspective. He then taught social anthropology at Sussex University, UK, for more than 20 years and, until his passing, was Professor Emeritus of Language in Education at King's College, London and Visiting Professor of Education at the University of Pennsylvania. He worked and lectured in the USA, Australia, Brazil and South Africa, amongst other countries, applying cross-cultural perspectives to educational issues around literacy, numeracy, language and development. He was also involved, with Alan Rogers, in an international project Learning Empowerment through Training in Ethnographic Research (LETTER) and was President of the British Association for Literacy in Development which works in both literacy and numeracy across countries.

Shanah Mompoloki Suping is a senior lecturer in science education in the Department of Maths and Science Education, University of Botswana. He specializes in chemistry education, with emphasis on the application of technology in teaching, as well as the use of locally available materials and contexts to make learning enjoyable, relevant and accessible.

Fernanda Wanderer is a professor at Universidade Federal do Rio Grande do Sul (UFRGS). One of her research areas is mathematics education and curriculum. Her most recent book is *Educação Matemática, Jogos de Linguagem e Regulação*, published in 2014.

Keiko Yasukawa is an adult numeracy and literacy researcher and teacher educator at the University of Technology Sydney in Australia. She takes a critical perspective in her research to seek to inform policy and practices in adult numeracy and literacy. She has undertaken research in numeracy and literacy practices in workplaces, vocational education and training colleges and community settings. She is a member of the editorial team of *Literacy and Numeracy Studies: An International Journal in the Education and Training of Adults*.

FOREWORD

The origin of this volume lay in a conversation between Alan Rogers and Brian V. Street in 2009. Reviewing a visit to adult literacy programmes in Africa, Alan mentioned a discussion in Kampala with the commissioner for the national adult literacy programme, during which the commissioner said: 'We don't really need any more literacy for development programmes: what we need is a *numeracy* for development programme'.

This was a theme which was becoming more pressing with the spread of mobile communication technologies throughout Africa, reaching many remote settlements and often used for texting by women and men who were inscribed (and who inscribed themselves) as illiterate. In 2005, Alan and Brian had organised an international seminar on numeracy (Uppingham Seminars 2005) and Brian had contributed to a Nuffield-funded project on home-school links for numeracy (Street et al. 2005). Brian said that he hoped such a programme, if it were ever created, would adopt a social practice view of numeracy using ethnographic-style research approaches, and Alan suggested that a publication similar to the *Social Uses of Literacy* (Prinsloo and Breier 1999), which had had a major impact on the field of adult literacy (the new literacy studies, now literacy as social practice), both academically and in practice, was needed to get numeracy onto the international development agenda.

So a proposal for an international publication on numeracy as social practice (NSP) was born. At the suggestion of Dr Stephen Black, Dr Keiko Yasukawa of University of Technology Sydney agreed to become a co-editor, as did Dr Kara Jackson, a former student of Brian, who came to that seminar from USA. A team of people engaged in direct research, especially some who were working in ethnographic-style research in so-called developing country contexts, agreed to contribute. It has not been an easy ride; one or two of the original authors found they could not keep their commitments and alternatives were found; and there have been other delays. The biggest blow was the death in June 2017 of Brian V. Street after a long fight against cancer. He retained an interest in the volume to the end; Alan was able to discuss it with him in some detail not much more than a week before he died.

The work has been shared out between us and the contributors have been generous with their time and expertise – we are very grateful to them. We wish to thank Dr Steve Reder who supported this project from a very early stage. We are also grateful to the team at Routledge who accepted that the original timetable would have to be extended in view of the editorial changes. And we are especially grateful to Maria Hays, a doctoral student at the University of Washington who helped with editing manuscripts and collating final versions of all the manuscripts; her meticulous attention to detail and ability to make sense of all our requests and comments were invaluable.

We hope this volume will illustrate why it is imperative that numeracy is treated as social practice, and that it will be a resource for reflection and innovation for practitioners, lead to further research, and create debates in pedagogy, research and policy.

Alan Rogers, Keiko Yasukawa and Kara Jackson
November 2017

References

Prinsloo, M. and Breier, M. eds. (1999) *Social Uses of Literacy* Amsterdam: John Benjamins

Street, B. V., Baker, D. and Tomlin, A. (2005) *Navigating Numeracies: Home/School Numeracy Practices* Dordrecht: Springer

Uppingham Seminars (2005) "Numeracy and Development" 20–22 October. Retrieved 18 January 2018 from http://www.uppinghamseminars.co.uk/report2005.pdf

Introduction

1

MAPPING THE TERRAIN OF SOCIAL PRACTICE PERSPECTIVES OF NUMERACY

Keiko Yasukawa, Kara Jackson, Phil Kane and Diana Coben

Introduction

How people think, act and reflect on the mathematical demands of a task is typical of what is called numeracy. At the same time, however, numeracy has become 'a rather elastic term' (Ainley and Doig 2001, 3), and its various uses in research, policy and practice suggest it remains 'a notoriously slippery concept' (Coben et al. 2003, 393). Dictionary definitions and some popular understandings of numeracy may still be limited to the notion of 'basic skills with numbers', with the assumption that there is some universality in the way these basic skills are applied and are accorded value irrespective of social and cultural context.

However, within research communities the term numeracy has a history of evolution from this 'basic skills' notion to one that focuses on what people do with mathematics, one that acknowledges its contingency to context. For example, based on their survey of the literature, Geiger, Goos and Forgasz (2015) define numeracy as the term

> used to identify the knowledge and capabilities required to accommodate the mathematical demands of private and public life and to participate in society as informed, reflective, and contributing citizens.
>
> *(Geiger et al. 2015, 531)*

Moreover, being numerate

> extends beyond the mastery of basic arithmetic skills to how to connect the mathematics learnt in formal situations, such as school classrooms, to real world problems....[It] involves the capability to: make sense of

non-mathematical contexts through a mathematical lens; exercise critical judgement; and explore and bring to resolution real world problems.

(Geiger et al. 2015, 531)

In this conceptualization, numeracy is a resource, not only to be a successful school student, but a resource central to negotiating the demands of life in multiple domains. While this definition of numeracy was developed through research in the school education sector with a view to informing curricula in these sectors, there is clearly scope for this model to be understood in lifelong and lifewide learning contexts. In fact, the Organization of Economic Co-operation and Development (OECD) Programme of International Assessment of Adult Competence (PIAAC) (OECD 2016, 18) defines numeracy as:

> the ability to access, use, interpret and communicate mathematical information and ideas in order to engage in and manage the mathematical demands of a range of situations in adult life. To this end, numeracy involves managing a situation or solving a problem in a real context, by responding to mathematical content/information/ideas represented in multiple ways.

This informed the Numeracy component of the PIAAC *Survey of Adult Skills* (OECD 2016).

While there has been progress in terms of how numeracy is conceptualized in some contexts, the ways in which numeracy teaching and learning is enacted often reflects and reinforces narrow conceptions of what constitutes 'numeracy' (see for example, Baker 1998, Coben et al. 2003). The chapters in this volume aim to highlight the importance of broadening our vision of numeracy in relation to teaching and learning and educational policy. All reflect what we refer to as a *numeracy as social practice perspective*.

In this chapter, we elaborate on what is meant by a perspective of numeracy as social practice (NSP). This perspective is influenced by several theoretical traditions which provide different analytical resources for raising and answering different kinds of research questions, rendering the terrain of NSP research sometimes difficult to navigate. While a diversity of perspectives enriches the broad numeracy research project, the sum of their contributions can only be viable as a resource for informing pedagogies and policy if the research terrain is mapped and signposted to indicate what could be gained from viewing numeracy through each of these different theoretical lenses. It is, therefore, imperative upon researchers of numeracy as social practice to clearly articulate the distinctive contributions their research perspectives can make to broaden debates about improving the numeracy of children and adults, and how this can be achieved.

In what follows, we introduce a selection of theoretical lenses that have been used to study numeracy as social practice, and in doing so, we identify the kinds of questions that these theoretical lenses have helped to address in research studies. In addition, we discuss some of their common and distinctive orientations, and consider how their contributions to NSP studies can be represented in relationship to each other.

Theoretical influences on numeracy as social practice

In our survey of the literature, we identified four different, though overlapping, theoretical models/schools of thought that inform studies of NSP: situative perspective on cognition; cultural-historical activity theory; new literacy studies or literacy as social practice; and ethnomathematics. This selection is not exhaustive of the theoretical resources that can inform the study of NSP; however, in our view, these four families highlight sets of assumptions that are central to taking a social practices perspective on numeracy. Although drawing on numeracy research that represents formal education, workplace studies and 'informal' learning, including both voluntary and involuntary learning, here we highlight contexts aside from compulsory schooling because research that has been concerned with broadening notions of what counts as numeracy has largely focused on other contexts.

Situative perspective on cognition

One major influence on viewing numeracy as social practice is the substantial body of research known as situative perspective on cognition. This grew out of research conducted in the 1970s to 1990s by psychologists (e.g., Nunes, Schliemann and Saxe) and anthropologists (e.g., Lave). This perspective was broadly framed as a critique of learning theories that privileged the notion of transfer, that is, that knowledge is viewed as 'tools for thinking' (Packer 2001, 498) that the individual transfers to different situations and applies to different tasks. According to those theories of transfer, the tools do not change and are 'independent of the situations in which they are used' (498). Knowledge is presumed to be 'context-free' or, in the words of Packer, 'detached from space and time, from specific contexts, from concrete experiences' (498).

In contrast, a number of scholars carried out studies of 'cognition in the wild' (Hutchins 1995) in which they showed that 'cognitive' activities are not separate from the settings and activities in which they take place. Notably, many of these studies focused on mathematics. This was not accidental. Much psychological research on learning (especially research associated with the study of transfer) had centered on people's application of mathematical procedures and skills because mathematics, as a field, was presumed to be generalizable knowledge and culture-free. However, studies of 'cognition in the wild' showed that the use and application of mathematical knowledge and skills were not independent of context.

For example, Nunes, Carraher and Schliemann (Carraher et al. 1985, Nunes et al. 1993) inquired into the numeracy practices of children who were engaged in candy-selling on the streets of Brazil. Candy-sellers conducted transactions that relied on sophisticated mental calculations with ease. However, when they were asked to perform those same transactions in a school-based manner, they were unable to do so. They abandoned their mental strategies (reflective of deep number sense) and struggled to solve the problems with school-based algorithms written down on paper. Studies like this showed that the context in which the mathematical activity was taking place was central to the children's facility with the mathematics, that competence in using mathematics was situational; these candy-sellers could easily have been viewed as deficient in computation in the school context, however they were clearly proficient in computation in the candy-selling context.

Saxe (1991, 2004) also investigated the mathematical practices of Brazilian candy-sellers. He was particularly taken by the candy-sellers' understandings of ratio given that they would have to determine the retail price of their candy in relation to the value of the continuously fluctuating Brazilian cruzeiro and the market price of the goods. Most of the sellers could not read the numbers on the notes and did not have any school-based knowledge of arithmetic algorithms. Saxe found that the sellers relied on grouping candies in relation to the size of the original box to determine the price of the candies. Saxe's work suggests that such quantification practices are 'socially and historically situated, constituted as individuals draw upon cultural forms (and the functions they afford) that themselves have complex social histories' (Saxe 2004, 245). Similarly, Saraswathi (n.d.) documented a range of counting and measurement practices in Tamil Nadu in India, including a way of counting cigarettes in bundles of 25, the number that fits in one handful when arranged in a particular way. In Saxe's and Saraswathi's studies, we see evidence of Lave's observation that close study of people's use of mathematics in contexts (especially those outside of formal schooling) reveals that mathematics itself is not static or devoid of context. '[T]he math[ematics] observed appears to have a *generative* relation with ongoing activities and at the same time to be *shaped* by them' (Lave 1988, 68, italics added).

Studies like these laid the foundation for establishing a situative perspective on cognition, in which '"[c]ognition" observed in everyday practice is distributed – stretched over, not divided among – mind, body, activity and culturally organized settings (which include other actors)' (Lave 1988, 1) and learning is represented by change in participation in an activity setting over time (Lave and Wenger 1991).

A key concept in a situated perspective is 'practice.' Theories of practice are rooted in the work of several scholars, including Marx, Giddens and Bourdieu. While different theorists disagree on particular points, in general, theories of practice seek to overcome a subject/object dualism (e.g., Bourdieu 1977, Lave 1988) Sociologically speaking, practice theory assumes that it is *in practice* that individuals shape and are shaped by systems. By practice, Lave (1988) refers to 'everyday activity,' which she defines as 'what people do in daily, weekly, ordinary cycles of activities' (15). Lave writes,

> It is the routine character of activity, rich expectations generated over time about its shape, and settings designed for those activities and organized by them, that form the class of events which constitutes an object of analysis in theories of practice.
>
> *(Lave 1988, 15)*

Practices involve cultural and social norms of participation, a relatively routine set of activities, and material artifacts, and typically involve people who are both 'novices' and 'experts' in that particular practice (Lave and Wenger, 1991). These norms, routines and the use of material artifacts change over time in relation to the change in individuals who participate in the activity and in their abilities to respond to new challenges.

Artifacts are an especially significant part of the context because they afford and constrain what can be accomplished and by whom (Brown et al. 1989, Packer and Goicoecha 2000, Lave 1988, Lave and Wenger 1991). For example, in the work of Saxe (2004) described above, grouping the candy in boxes afforded an approach to determining cost based on proportional reasoning. If the boxes had been different sizes or if the candies had not been stored in boxes, the approach to determining cost might have been quite different. In Nabi's (2009) study, a Pakistani bangle-seller hired an 'educated person' (90) to improve and computerize his shop's record-keeping system. The hired man went about installing a new system, refusing to understand the significance of the shop's system of naming and the measurements, with a result where:

> the whole shop became a big mess and nobody accepted these changes to the system. Our language is built into the glass bangles industry and cannot be separated from it. So he had spoiled the management of the shop, and I had to close the shop for three days.
>
> *(Nabi 2009, 90)*

As Wertsch (1998) argues, artifacts can be both material, such as a calculator and immaterial, such as language (see Pahl and Rowsell 2011 on artifactual literacies; Holland, Lachicotte, Skinner and Cain 1998 on figured worlds) and both contribute to the semiotic mediation of human action. Immaterial resources are ubiquitous in mathematics. Wertsch (1998) illustrates this with a description of multiplication where many people draw on a familiar algorithm and subsequently align the numbers in a vertical fashion, and then perform a series of multiplication problems, that when added, result in the correct solution. Although this algorithm is a cultural resource, it is largely invisible. Furthermore, Wertsch uses this example to argue that practically any mathematical task is *jointly accomplished between an individual and a tool.*

Cultural-historical activity theory (CHAT)

Another influence on numeracy as social practice perspective is cultural-historical activity theory (CHAT). CHAT provides a lens that has the potential to examine numeracy practices systematically, particularly in situations where historically well-established practices experience a 'disturbance' and are faced with change.

The concept of activity was first introduced by Vygotsky to explain learning and concept formation as a goal-oriented and mediated activity (Vygotsky 1986). More recently, a number of researchers have looked to Engeström's (2001) formulation of the third-generation activity theory (the second generation being attributed to Leontiev), particularly in researching numeracy and mathematical practices in workplaces (Fitzsimons 2005, Kanes 2002, Pozzi et al. 1998, Triantafillou and Potari 2010, Roth 2012, Williams and Wake 2007).

For Engeström (1993), the notion of an activity system enables us to think about 'context' in terms of the historical, social and dynamic arena in which collective practices, including numeracy practices, emerge, are reproduced and evolve. In Engeström's formulation, an activity system is constituted by the *subject(s)* from whose perspective the activity system is perceived and experienced, the *object* (purpose or motivation of the activity), the *mediating artifacts* (symbolic and material tools and instruments) used by the subject, the *rules* (formal and informal) of the *community* in which the activity system is embedded, and the *division of labour* within that community (Engeström 1993). In this formulation, the activity system that embraces the collective unit in which the subject (either an individual or a group) is located, and the traditions and the power relations/hierarchies operating within and on the collective, become the unit of analysis. This focus on the activity system as collectively constituted renders this third-generation CHAT particularly useful in workplace numeracy research, but also more generally in understanding the production, continuity and transformation of numeracy practices in particular contexts.

Using CHAT, FitzSimons (2005) conducted research in Australian workplaces, including: 'a fundraising "trivial challenge" production office, a modular shed construction company, a local post office, a short-term home rental company, a graphic design company using computer-controlled machinery, a local playgroup, a small hairdressing salon, a wholesale power tool warehouse, and an aged-care hostel' (30). Her research findings resonate with those from other research which found that in many workplace tasks:

> mathematical elements in workplaces are subsumed under workplace routines, structured by mediating artefacts (e.g., tools and equipment, calibration templates, record sheets), and are highly context-dependent. In other words, the priority is to get the job done as efficiently as possible — not to practise and refine mathematical skills.
>
> *(FitzSimons 2005, 37)*

On the other hand, the results of the estimation, calculations and measurements in these work tasks have real consequences, for example to safety, deadlines, cost and ensuring any competitive advantage.

Triantafillou and Potari (2010) focused closely on one element of the activity system in their research on the role of tools. Through an ethnographic study of telecommunications technicians out in the field, the researchers observed and developed a taxonomy of a wide range of tools being brought to play to accomplish mathematical tasks in the workplace: physical tools such as wires, telecommunications closets and measuring instruments; written texts such as manuals, technical notes and work protocols; communicative tools including diagrams, maps, metaphors, coded data, formulae, algorithms; and cognitive processes and concepts including calculations, data interpretations, measurements, and number system concepts and algebraic relations. While such a taxonomy is interesting for researchers, the workers did not necessarily see themselves as engaged in mathematics or numeracy.

Whereas a researcher interested in NSP may 'see' numeracy as the object of the workplace activity systems – e.g., solution to a telecommunications network problem – for the members of the workplace, the object is rarely, if ever, that. The numeracy is not visible, but useable (Kanes 2002) in producing the outputs of the task at hand. In a study of production workers' numeracy (and literacy) practices, Yasukawa, Brown, and Black (2013) saw complex, technology-mediated numeracy practices in the production of customized hearing aids which were downplayed by the workers as 'just like playing a video game' (377). For the subject of the workplace activity system, the object is certainly not numeracy; it is their daily production quota of hearing aids that they need in order to help their team meet the daily target.

CHAT's interest in expansive learning – the radical transformation of an existing activity system as a result of some 'disturbance' – enables insights into the historical continuity and discontinuity of numeracy practices. In Yasukawa et al.'s (2013) study of the production workers, the company had recently purchased a three-dimensional printer to 'print' the shells of the customized hearing aids. The printer introduced a disturbance to their historical practice of manually molding the shells. Workers now had to model the shells on their computers using software with a graphic interface that enabled them to view projections of the shell during its formation through scaling and rotating the image on the screen. Although modelling using the graphic software was new to them, the workers indicated that it was not difficult because they had done the 'same' work manually, before the printer was introduced; from their previous experience they brought a deep understanding of the aesthetics, technical requirements and comfort for the user. Thus, even though many of the mediating tools and instruments for producing the shells had changed dramatically, there were critical elements that the workers were able to 'transfer' into the newly configured activity system of production. The focus of CHAT on the historicity of activity systems invites the researcher to interrogate the history that the workers bring to their current practice, and note how changes in one set of elements in the activity system do not always lead to a radical reconceptualization of the work.

Literacy as social practice

Another influence on a numeracy as social practice perspective is what is known as literacy as social practice (LSP). LSP grew out of the work of anthropologist Brian V. Street, who studied the different ways in which villagers in Iran used literacy in their lives: in religious classes, schooling and commercial activities. On the basis of his ethnographic work, Street contrasted two different models to make sense of how literacy could be framed – autonomous and ideological models. He classified the autonomous model of literacy as working 'from the assumption that literacy in itself—autonomously—will have effects on other social and cognitive practices' (Street 2001, 7). This model acts as if literacy skills are independent of any particular ideology and presents these skills 'as neutral and universal' (7). In contrast, Street described ideological models of literacy as

> posit[ing] instead that literacy is a social practice, not simply a technical and neutral skill; that it is always embedded in socially constructed epistemological principles. ... Literacy in this sense is always contested, both its meanings and its practices, hence particular versions of it are always 'ideological', they are always rooted in a particular world-view and a desire for that view of literacy to dominate and to marginalize others.
>
> *(Street 2001, 7–8)*

In this respect, then, 'it is not valid to suggest that "literacy" can be "given" neutrally and then its "social" effects only experienced afterwards' (8). Street's approach to understanding literacy became known as new literacy studies (NLS), but as understandings of literacy as social practice continued to be debated as new social practices emerged alongside new technological and global influences, Street (2016) himself started to prefer LSP as the term to refer to contemporary studies of literacy as social practice.

Drawing on the ideas of NLS, Baker (1998) drew analogies to numeracy. He characterized an autonomous model of numeracy as one which 'presents numeracy as a set of pure skills separate from contexts in which they may be used, showing its belief that numeracy is both culture- and value-free' (37). Alternatively, an ideological model of numeracy 'sees numeracy as social practices with its content or body of knowledge sited within contextual, cultural and ideological circumstances' (38–39). The ideological model views numeracy, like literacy, as both the product of society's culture and cultural history, and as the basis of commerce, science, technology and other everyday transactions, all of which are located in power structures. An ideological model also highlights the power relations that are imbued in numeracy practices, including how some forms of numeracy knowledge and activity are accorded higher status over others.

Building on NLS's understanding of a literacy event, in which a piece of writing is central to the event (Heath 1982), Street, Baker and Tomlin take *numeracy events* to be 'occasions in which numeracy activity is integral to the nature of the participants' interactions and their interpretive processes' (Street et al. 2005, 20). Similarly, *numeracy practices* are taken to be:

> more than the behaviours that occur when people do mathematics/ numeracy – more than the events in which numerical activity is involved. [They enable the researcher] to explore the conceptualizations, the discourse, the values and beliefs and the social relations that surround numeracy events as well as the context in which they are located.
>
> *(Street et al. 2005, 20)*

As an example of scholarship that takes up a LSP perspective on numeracy, Anderson and Gold (Anderson and Gold 2006, Gold and Mordecai-Phillips 2003) followed four low-income African American pre-kindergarten children across home and school contexts in a US city. They documented 'travel,' when an individual tries to enact a similar mathematical strategy in different contexts. Drawing on the numeracy studies framework put forth by Street et al. (2005), they found that strategies travel between sites but are often given different meanings in these different contexts. For example, a five-year-old playing 'chutes (snakes) and ladders' in the home demonstrated sophisticated one-to-one correspondence. He projected the number of spaces he needed to reach a ladder and then, depending on his roll, would tap the board the correct amount of times; in doing so, he often had to double-count a space in order to reach the ladder. He was allowed to do this in the home and was considered intelligent by his grandmother for doing so, but when he enacted this same strategy in the school, he was labelled a cheater. The teacher did not recognize that his strategy was actually quite sophisticated for a five-year-old and reflected number sense.

Ethnomathematics

A fourth influence on a numeracy as social practice perspective is ethnomathematics. Ethnomathematics arose as a field of study in the 1980s in light of concern about the power relations between what is legitimated in mathematics textbooks and the traditional everyday mathematical practices of different cultures. In his seminal 1985 paper, D'Ambrosio writes:

> we call ethno*mathematics* the mathematics which is practised among identifiable cultural groups such as national-tribal societies, labor groups, children of certain age bracket, professional classes, and so on. Its identity

depends largely on focuses of interest, on motivation, and on certain codes and jargons which do not belong to the realm of academic mathematics. We may go even further in this concept of ethnomathematics to include much of the mathematics which is currently practised by engineers, mainly calculus, which does not respond to the concept of rigor and formalism developed in academic courses of calculus.

(D'Ambrosio 1985, 45)

D'Ambrosio writes that it is a research field that extends across history and cultural anthropology, interrogating the history and nature of mathematics (D'Ambrosio 1985, 2001). In addition, Gerdes (1994) talks about the *ethnomathematical movement*: a research field that aims to influence mathematics education and curriculum to be more empowering and emancipatory for all learners – a vision shared by other researchers including D'Ambrosio (1985, 1997, 2001), Borba (1990), Ascher (1991), Knijnik (1999), and Zaslavsky (1994). These researchers of ethnomathematics are not concerned simply with uncovering a diversity of mathematical practices or, to use Gerdes' (1994) term, to 'unfreeze' mathematics that has been frozen in old techniques and rendered invisible.

D'Ambrosio (1985) makes explicit that ethnomathematics cannot avoid the role of ideologies in the epistemological legitimation afforded to Western mathematics, and the struggles surrounding the foregrounding of ethnomathematical ideas in 'mainstream' educational discourses. The ethnomathematics movement is underpinned by a strong ethical, social justice stance to revitalize the epistemological assumptions about mathematics, and in turn to revitalize the mathematics curriculum:

When ethnomathematicians say 'more than one mathematics', they are recognizing different responses to different natural, social, cultural environments. ... there may be more than one way of explaining, understanding, coping with reality. ... I see a possibility of new civilization, through elimination of inequity, bigotry, intolerance, hatred, discrimination, as the result of unfreezing the forms of knowledge ... and allowing cultural dynamics to play its role in the evolution of species.

(D'Ambrosio, 2001, 68)

However, ethnomathematics researchers and activists are under no illusion that this is straightforward, they do not see these aims achieved simply by 'adding' ethnomathematical findings in an ad hoc manner into the existing curriculum. For example, Knijnik (2002) outlines the challenges of an ethnomathematics-based pedagogy she has been developing in her work with the landless people's movement in Brazil as one of constant tension of working with the popular knowledge of the landless people and introducing 'academic' knowledge of mathematics in culturally and politically nuanced ways.

Borba (1990) and Zaslavsky (1994) both provide pedagogical directions for incorporating ethnomathematics in holistic ways into the mathematics curriculum. Borba draws on Freirean notions of dialogue and the generation of authentic problems in the learners' own lives as the starting point. Zaslavsky suggests giving learners 'the opportunity to see the relevance of mathematics to their own lives and to their community, to research their own ethnomathematics' (Zaslavsky 1994, 6), and that by 'taking into account their out-of-school experience, mathematics should help students to pinpoint and take action of the societal factors that stand in the way of their living fulfilled lives' (7). Borba (1990) observes, however, that much of the efforts and successes of such approaches have been in the adult and non-formal educational sectors, and much less so in the formal school education sectors, except through some initiatives using themes and cross-disciplinary project work. He suggests that the constraints of the traditional curricula, not just in mathematics, but in other subjects as well, need to be challenged in order for the ethnomathematics project to be the holistic approach of allowing students' lives to be part of the learning.

Key features of a numeracy as social practice (NSP) perspective

Whereas the origins of a numeracy as social practice perspective are multiple, there are several key features that, in our view, distinguish this perspective.

A NSP perspective focuses on what people *do* with numeracy through social interactions in particular contexts, rather than on people's performance of mathematical skills in isolation of context. As all four lenses described above highlight, the forms of activity in which people engage that entail numeracy are situated in context. Moreover, a focus on *practice* entails viewing numeracy activity as culturally, historically and politically situated. Indeed, *all numeracy activity is ideological*.

Another feature of a social practice perspective is the interest in both *visible* and *invisible* mathematics. Many studies using a social practice lens have uncovered practices that are easily overlooked as mathematics or numeracy practices, both by the 'doers' and researchers of numeracy. And, as all four lenses indicate, this means we need as a field to contend seriously with the power relations that are entailed in teaching, learning and researching numeracy practice. For example, power relations are in play when a proficient numeracy user is reluctant to share information in case it threatens their own position. 'Protecting their own patch' could be viewed as a defensive social practice. This relates to what is made visible and counted as 'numeracy' or 'mathematics' and whose practices are privileged in accounts of numeracy.

The theoretical lenses described above highlight particular aspects of a NSP perspective, and offer specific analytical resources to investigate teaching and learning activity from this perspective. For example, how the local context shapes social practices, including numeracy practices, is one of the key questions addressed by theories of cognition that reflect a situative perspective, as well as by LSP. Like situated cognition theories, a CHAT formulation of numeracy practices is focused on the goal-oriented doing of mathematics by the subjects. However, CHAT also enables the researcher to focus on how particular numeracy practices have been shaped or disrupted by rules and traditions, the mediating tools and instruments available, and the community in which the numeracy practices have meaning and value.

The concerns of ethnomathematics researchers resonate with those of the LSP-informed researchers of NSP. Both highlight tensions between Western/academic mathematics and ethnomathematics of particular cultural groups. Both perspectives make clear the inseparability of culture and the mathematics that people do, and that there are historical continuities and discontinuities of mathematical practices that reflect different forms of power being exercised. And neither LSP nor the ethnomathematics researchers are satisfied with uncovering 'exotic' forms of mathematical activities – there are broader educational agendas motivating the work. There are also saliences with CHAT, viewing on the one hand the ethnomathematics of different groups as distinct cultural-historical activity systems whose objects are related to the authentic problems and situations of the group in a particular place and time, and on the other, the 'academic mathematics' whose object is arguably the universal education of a standard curriculum.

Finally, all four lenses necessarily draw on ethnographic approaches to studying the numeracy practices *in-situ*. Such approaches enable thick descriptions and voices of those participating in the numeracy practices to be foregrounded. Being case studies, the particularities of the findings do not in themselves generalize to other contexts; however, they illuminate the gaps that necessarily exist in large scale survey-based research, standardized assessments and off-site interview-based research with people who comment on others' numeracy. Importantly they challenge in very powerful ways popular deficit discourses that make blanket statements about poor performance of entire populations, and equally ill-informed 'back to basic' mandates to 'fix' these deficits. NSP does not discount the importance of skills or school-based learning, but what it does show is that skills and knowledge of mathematics that are devoid of context do not enable people to be productive and accepted members of particular communities of practice. In any specific community of practice, these skills and knowledge are embedded into and transformed into activities that have particular meanings for the members of the community.

Preview of forthcoming chapters

The chapters that follow each illustrate the importance of understanding numeracy as social practice, particularly in the contexts of lifelong and lifewide learning. An important aspect of this volume is that it deliberately includes attent on to diversity along a number of dimensions. One dimension concerns the geographical contexts they write about; we deliberately sought out an international set of authors, including from both developed and developing societies. A second dimension concerns the contexts that the authors research, including informal education, formal education, the workplace, as well as livelihood occupations and everyday life. A third dimension concerns the different theoretical resources the authors bring to researching numeracy as social practice. Finally, we have deliberately sought to include both established researchers and some 'younger' voices.

The forthcoming chapters in this volume are organized into parts in order to highlight the diversity of numeracy practices, and what researching numeracy practices entails given this diversity (Part I); the theoretical resources that researchers of NSP use to uncover the sociocultural factors in the interplay between everyday numeracy practices and formal mathematics education (Part II); and the political dimensions of numeracy and mathematics learning that a NSP approach can reveal (Part III). Thus:

- Part I includes a number of case studies of numeracy practices. Each chapter explores the relationship between these numeracy practices and the more formal mathematics of the classroom and/or training center.
- Part II illustrates the different resources that can be drawn upon to study interactions (or lack thereof) between mathematics education and people's everyday numeracies. These studies make significant contributions not only to NSP research, but numeracy and mathematics education more broadly given the widespread nature of the concerns being addressed.
- Part III highlights in particular the crucial role that power plays in making sense of what is or can be learned numeracy-wise, by whom, and for what purposes. The authors focus on a range of learning contexts, including compulsory education, adult education and workplace settings.

Each is prefaced by an introduction that provides an overview of the studies within that Part.

The chapters are followed by a Conclusion in which the editors revisit the terrain of numeracy as social practice described in this chapter in light of the contributions made by the authors in the volume. The editors first discuss some emergent themes and draw out implications for practice and policy.

References

Ainley, J. and Doig, B. (2001) *Summing Up: Australian Numeracy Performances, Practices, Programs and Possibilities* Melbourne: ACER Press.

Anderson, D. and Gold, E. (2006) "Home to school: Numeracy practices and mathematical identities" *Mathematical Thinking and Learning*, 8(3), 261–286.

Ascher, M. (1991) *Ethnomathematics: A Multicultural View of Mathematical Ideas* Pacific Grove, CA: Brooks/Cole.

Baker, D. (1998) "Numeracy as social practice" *Literacy and Numeracy Studies,* 8(1), 37–50.

Borta, M. C. (1990) "Ethnomathematics and education" *For the Learning of Mathematics*, 10(1), 39–43.

Bourdieu, P. (1977) *Outline of a Theory of Practice* (R. Nice, trans.) Cambridge: Cambridge University Press.

Brown, J. S., Collins, A., and Duguid, P. (1989) "Situated cognition and the culture of learning" *Educational Researcher*, 18(1), 32–42.

Carraher, T. N., Carraher, D. and Schliemann, A. D. (1985) "Mathematics in the streets and in schools" *British Journal of Developmental Psychology*, 3(1), 21–29.

Coben, D., Colwell, D., Macrae, S., Boaler, J., Brown, M. and Rhodes, V. (2003) *Adult Numeracy: Review of research and related literature* Retrieved 18 January 2018 from http://dera.ioe.ac.uk/22487/1/doc_2802.pdf.

D'Ambrosio, U. (1985) "Ethnomathematics and its place in the history and pedagogy of mathematics" *For the Learning of Mathematics*, 5(1), 44–47.

D'Ambrosio, U. (1997) "Where does ethnomathematics stand nowadays?" *For the Learning of Mathematics*, 17(2), 13–17.

D'Ambrosio, U. (2001) "General remarks on ethnomathematics" *ZDM Mathematics Education*, 33(3), 67–69.

Engeström, Y. (1993) "Work as a testbench of activity theory". In Chaiklin, S. and Lave, J. (eds), *Understanding Practice: Perspectives on Activity and Context* Cambridge: Cambridge University Press.

Engeström, Y. (2001) "Expansive learning at work: Toward an activity theoretical reconceptualization" *Journal of Education and Work*, 14(1), 133–156.

FitzSimons, G. E. (2005) "Numeracy and Australian workplaces: Findings and implications" *Australian Senior Mathematics Journal*, 19(2), 27–40.

Geiger, V., Goos, M. and Forgasz, H. (2015) "A rich interpretation of numeracy for the 21st century: A survey of the state of the field" *ZDM Mathematics Education*, 47, 531–548.

Gerdes, P. (1994) "Reflections on ethnomathematics" *For the Learning of Mathematics*, 14(2), 19–21.

Gold, E. and Mordecai-Phillips, R. (2003) "Learning in multiple worlds: An examination of the intersection of home and school mathematics practice" American Educational Research Association (AERA) Annual Meeting, 21–25 April, Chicago, IL.

Heath, S. B. (1982) "What no bedtime story means: Narrative skills at home and school" *Language and Society*, 11, 49–76.

Holland, D., Lachicotte, W., Skinner, D. and Cain, C. (1998) *Identity and Agency in Cultural Worlds* Cambridge, MA: Harvard University Press.

Hutchins, E. (1995) *Cognition in the Wild* Cambridge, MA: MIT Press.

Kanes, C. (2002) "Towards numeracy as a cultural historical activity system". In Valero, P. and Skovsmose, O. (eds), *Proceedings of the 3rd International MES Conference* Copenhagen: Centre for Research in Learning Mathematics.

Knijnik, G. (2002) "Curriculum, culture and ethnomathematics: The practices of 'cubagem of wood' in the Brazilian landless movement" *Journal of Intercultural Studies*, 23(2), 149–165.

Lave, J. (1988) *Cognition in Practice: Mind, Mathematics and Culture in Everyday Life* Cambridge: Cambridge University Press.

Lave, J. and Wenger, E. (1991) *Situated Learning: Legitimate Peripheral Participation* Cambridge: Cambridge University Press.

Nabi, R. (2009) "Sajid, the glass bangle seller, runs rings round mental arithmetic". In Nabi, R, Street, B, and Rogers, A. (eds), *Hidden Literacies*, Bury St Edmunds: Uppingham Press.

Nunes, T., Schliemann, A. D. and Carraher, D. (1993) *Street Mathematics and School Mathematics* Cambridge: Cambridge University Press.

OECD (2016) *The Survey of Adult Skills: Reader's Companion,* second edition, Paris: OECD Publishing.

Packer, M. J. (2001) "The problem of transfer, and the sociocultural critique of schooling" *Journal of the Learning Sciences*, 10(4), 493–514.

Packer, M. J. and Goicoecha, J. (2000) "Sociocultural and constructivist theories of learning: Ontology, not just epistemology" *Educational Psychologist*, 35, 227–241.

Pahl, K. and Rowsell, J. (2011) "Artifactual critical literacy: A new perspective for literacy education" *Berkeley Review of Education*, 2(2), 129–151.

Pozzi, S., Noss, R., and Hoyles, C. (1998) "Tools in practice, mathematics in use" *Educational Studies in Mathematics*, 36(2), 105–122.

Roth, W. M. (2012) "Cultural-historical activity theory: Vygotsky's forgotten and suppressed legacy and its implication for mathematics education" *Mathematics Education Research Journal*, 24(1), 87–104.

Saraswathi, L. S. (n.d.) *Everyday Mathematics and the Classroom: Case Studies from Rural South India*. Retrieved 18 January 2018 from http://www.balid.org.uk/wp-content/uploads/2012/10/Saraswathi-Everyday-mathematics-and-the-classroom.pdf.

Saxe, G. B. (1991) *Culture and Cognitive Development: Studies in Mathematical Understanding* Hillsdale, NJ: Lawrence Erlbaum.

Saxe, G. B. (2004) "Practices of quantification from a socio-cultural perspective". In Demetriou, K. A. and Raftopoulos, A. (eds), *Developmental Change: Theories, Models, and Measurement* New York: Cambridge University Press.

Street, B. V. (2001). *Literacy and Development: Ethnographic Perspectives* London: Routledge.

Street, B. V. (2016) "Literacy and development: Ethnographic perspectives" presented at Headquarters of the National Union of Teachers, Hamilton House, London, April 26–27. Retrieved 18 Janaury 2018 from http://www.ramphalinstitute.org/uploads/2/3/9/9/23993131/literacy_and_development_by_prof_b_street.pdf.

Street, B. V., Baker, D. and Tomlin, A. (2005) *Navigating Numeracies: Home/School Numeracy Practices* Dordrecht: Springer.

Triantafillou, C. and Potari, D. (2010) "Mathematical practices in a technological workplace: The role of tools" *Educational Studies in Mathematics*, 74(3), 275–294.

Vygotsky, L. (1986) *Thought and Language*, revised edition (A. Kozulin, trans.) Cambridge, MA: MIT Press

Wertsch, J. V. (1998) *Mind as Action* New York: Oxford University Press.

Williams, J. and Wake, G. (2007) "Black boxes in workplace mathematics" *Educational Studies in Mathematics*, 64(3), 317–343.

Yasukawa, K., Brown, T. and Black, S. (2013) "Production workers' literacy and numeracy practices: Using cultural-historical activity theory (CHAT) as an analytical tool" *Journal of Vocational Education & Training*, 65(3), 369–384.

Zaslavsky, C. (1994) "'Africa counts' and ethnomathematics" *For the Learning of Mathematics*, 14(2), 3–8.

PART I

Using case studies to expose the significance of what 'surrounds' mathematics in numeracy practices

Alan Rogers and Brian V. Street

Numeracy as social practice (NSP) understands numeracy always and only to be practised in a specific context on a specific occasion. The first Part of this collection of studies includes four research projects – 'stories from the field' – which examine different activities, exploring what surrounds the numeracy practices, what it is that gives them meaning. They come from very different parts of the world. All reveal how numeracy practices are used in combination with other social and cultural practices to achieve specific goals.

In Chapter 2, Kane discusses two examples, how waste collection vehicle drivers work out each day the most effective ways to deliver their services and how the managers and workers on a fruit orchard in New Zealand determine the practices they need throughout the season to maximize the yield of the orchard. His examples point to the way that a fine balance between more or less precise measurements and estimations was used in the everyday numeracy practices. An 'audit' of the mathematical skills involved would show, for example, that the orchard managers needed to be able to measure accurately the distance between the residual canes of the kiwifruit vines, and to read temperature measurements to know when to commence pest monitoring. For the waste collectors, there were measurements of weight (the weight of the bins) and time (when they can start collecting, when they can get a smooth run, etc). They also needed to have a developed spatial awareness – of angles and distances to effectively manoeuvre the hydraulic arm to pick up the bins. However, these work practices also called for judgements to be made on the basis of long experience and local knowledge, for example, the weather patterns, and local streetscape and traffic patterns which interact with the mathematics embedded in these tasks.

In Chapter 3, Alangui admires the efficiency of those who build and repair stable and lasting stone walls to support their rice farming terraces in the Philippines. The mathematical knowledge concerns space and shapes, selecting the stones which best corresponded in weight and shape to the spaces which needed filling, ordering or classifying them according to size and shape, and rotating and positioning them. Wisdom from long experience and community tradition joined with (largely unconscious) mathematical practices to form the numeracy practices which helped them to achieve their purposes. They did not always get it right – some walls fell; but repairs reinforced the practices.

In Chapter 4, Kalman and Solares join Mexican workers checking their wages and ensuring they are not exploited by the traders they deal with. Power relations springing from long historical currents help format the numeracy practices involved – number recognition, counting and calculations, not just numerical but also of value, calling for quick decisions.

In Chapter 5, Boistrup and her colleagues analyse the way in which students in Sweden engaged in building a garden as part of the practical work for their courses use different numeracy practices and relate these to the mathematics they are learning in their formal education courses. More or less accurate measurements, more formal formulae (such as Pythagorean triangles) and symmetry in the patterning of the tiles being laid, and calculations of building material needed were made in the light of the available resources.

Within each of these examples, mathematical knowledge and skills joined with local knowledge and skills formed the practices that enabled goals to be achieved.

2

ESTIMATION BY KIWIFRUIT ORCHARD MANAGERS AND URBAN REFUSE/RECYCLING OPERATORS WITHIN THEIR SITUATED HORTICULTURAL OR CIVIC WORKPLACE PRACTICES

Case studies from New Zealand

Phil Kane

Introduction

Researchers have investigated people's workplace mathematics in occupations such as telecommunications technicians (Triantafillou and Potari 2010), taxi-drivers (Chase 1983), and boat-builders (Zevenbergen and Zevenbergen 2004). Each study explores how people make sense of quantities and workspaces in their roles, although the mathematics used is usually within a wider set of situated (Lave 1993) everyday practices. According to Lave (1996), as people make sense of their circumstances, they are constructing their own *in situ* identities. Situated learning cannot therefore be passive (Stein 1998); rather action and learning take place connecting people, locations, processes, contexts and situations.

In this chapter I draw from a social practices framework (Street 1995, Baker 1998), specifically from a new literacy studies (NLS) perspective (Gee 1996), to characterize practices entailing numeracy in which orchard managers and refuse/recycling operators engaged. This is counter to the traditional autonomous model of literacy (Street 2012) where reading and writing are valued over oracy and traditional forms of communication. A social practices model of literacy is not reliant on discrete skills but focuses instead on 'social, cultural, historical, and institutional contexts' (Gee 2010, 5). Barton (2006, 22) defines 'what people do *with literacy*' as 'literacy practices … However, practices are not observable units of behaviour since they also involve values, attitudes, feelings, and social relationships'. The social practices model is perhaps a closer fit to literacy than to numeracy. When viewing an incorrect decimal point in a bank transfer, or an erroneous ratio of raw materials in an industrial process, the implications of such errors have greater significance (Cockcroft 1982).

A case study research design was used in each workplace since 'the variables were so embedded in [each] situation as to be impossible to identify ahead of time' (Merriam 2009, 45). A single-case study approach was employed since participants' efforts typically model their real-time everyday jobs (Yin 2009). In the first case, orchard managers occupy a major link in the production chain, growing export-quality kiwifruit in each site they oversee. Their efforts are ultimately judged on post-harvest appearance, freshness and consumer taste of the fruit, and returns for grower/owners, or 'orchard gate returns' (NZKGI 2016). In the second case, refuse/recycling operators work individually in mobile settings and are responsible for emptying every bin in their territory. Operators must comply with the rules of their organizations and the regulations of the destination depots where they unload. They are regulated also by a national road transport authority and the local territorial council authority with its occasionally disgruntled residents. The perspective of numeracy as situated social practices is appropriate since the participants' numeracy in each context is not a set of isolated skills, rather an authentic, complex, varied and meaningful part in an organization's milieu. Each participant contends with commercial and/or civic interests and authority.

The following section begins with a discussion of estimation. Backgrounds of the two workplaces and the participants are then described (all names are pseudonyms), before their respective work practices and instances where estimation is used are given.

Estimation

Estimation practices are of interest for two reasons. First, there are many instances of estimation by orchard managers and refuse/recycling operators that align with the key mathematical competences across many workplaces (Hodgen and Marks 2013). Second, estimation traverses other mathematical themes in workplaces (Sowder 1992) such as number sense, measurement and spatial reasoning. Estimation is pervasive and probably extends beyond other fundamental quantitative abilities (Siegler and Booth 2005). Adept estimators know when to estimate, when to 'trade-off between simplification and proximity' (LeFevre, Greenham and Waheed 1993, 120), and they can detect an incorrect quantity or reading, then take remedial action. But it is impractical and time-consuming to measure or count everything, so at times estimators will 'sacrifice precision for convenience' (Gal 1999, 11). Over time estimates evolve to become benchmarks or landmarks, although some are more precise depending on the context (Hogan and Morony 2000).

Siegler and Booth (2005) define two categories of estimation. Numerical estimation is when an amount does not need to be exact, rather it is close to the correct magnitude. Sowder (1992, 373) describes 'clos[ing] in on a target value' as an approximation. Approximating enables decision-making but does not require more time spent calculating. LeFevre, Greenham and Waheed (1993) suggest time is better spent deciding on *when* to estimate, and on the concepts around number sense to provide a sound base for estimation. Non-numerical estimation (Siegler and Booth 2005) involves spatial and relevant geometrical reasoning other than calculating. Non-numerical estimations may be influenced by natural phenomena (e.g. temperature) or by idiosyncratic strategies. For instance, Adams and Harrell (2010, 12) describe a strategy about estimating tyre tread depths, where an American penny is placed in a tyre tread and if one saw 'all of Lincoln's head [then there was] less than 3/32 of an inch [so] your tire is worn out'. Smith (1999) adds that work teams train newcomers to industry standards, so practices (like estimation) take time to develop, especially as newcomers must become familiar with the quality ranges and, perhaps, any statistical process controls.

The workplaces and their participants

Kiwifruit orchard managers

After viewing horticultural students inspect kiwifruit vines at a polytechnic orchard, a case study of orchard managers was initiated. In view of the breadth of the New Zealand kiwifruit industry, the numeracy practices in the pivotal position of orchard managers only were explored. Three managers completing a horticultural qualification at the polytechnic were invited to participate and two, Gary and Dave, volunteered. Interviews revealed that orchard work was essentially seasonal, so this guided the coding of their stories. FitzSimons (2000) notes that with people's often complex workspaces, the mathematics they engage in is not always definitive. Accordingly, this case study focused on the numeracy practices drawn on by managers during their seasonal workplace practices.

An easily grown 'backyard fruit', the 'chinese gooseberry' (from 1959 'kiwifruit') has been grown commercially since the early 1940s. By 2007, over 2500 kiwifruit growers contributed 29 per cent of New Zealand's total horticultural exports (Mainland and Fisk 2006, Campbell and Haggerty 2008, HortResearch Rangahau Ahumara 2007), twice the volume of apples, the next largest export. The Bay of Plenty (eastern North Island) with its favourable climate and soils has become the prime commercial growing region in the country.

The managers had contrasting levels of experience: Gary had worked in the region since the 1980s at first in dairy farming. Following a serious accident, a chance encounter led him and his partner to kiwifruit. Thirty years on, the couple now manage orchards and teams of workers for several owner-growers in the Eastern Bay. The younger manager, Dave, was a recent arrival although like Gary he had first worked elsewhere; his cadetship with a packing company acquainted him with every facet of the industry. During his orchard initiation,

Dave was mentored by a senior manager, Rod, who had over twenty years' experience growing kiwifruit.

Recycling and refuse operators

Following a pilot investigation (Chana and Kane 2010), the second case study investigated collection operators of two South Auckland refuse and recyclables collection companies. Company X employed refuse/recyclables operators while Company Y contracted recyclables operators only. During the study, Company Y ended its Auckland operations owing to unrelated concerns offshore, and Company X successfully tendered for the vacant territories and absorbed Company Y's thirteen operators. These contractors joined the team of operators but each employment structure remained. By the end of the study, Company X had a combined team of about thirty-five operators, with a continuum of organizational responsibility, with owner-operators at one end and employed operators at the other. The numeracy practices of contractors are more complex; as with any small business, they must meet legal and financial obligations.

Information sheets inviting participation were distributed in each case, but the gate-keeping roles (Hennink, Hutter and Bailey 2011) of the horticulture tutor and the depot site managers respectively meant they circulated the invitations initially. Five collection operators volunteered, males aged 40 to 50. Three of the five, Dennis, Brian, and Sean, were Company X employees while Josh and Noel were originally Company Y contracted owner-operators. Sean had worked there for three months, while Dennis and Brian had worked there much longer. Participants could withdraw at any time, however most were curious to see why anyone might be interested in their work.

When the study began, the newly formed Auckland Council (representing 1.46 million residents) inherited refuse/recycling practices of seven former territorial local authorities, and merged the various kerbside collection practices. In 2010, refuse companies loaded 248,500 tonnes of kerbside waste (Middleton 2011, about 21 per cent of the total annual waste across the city with the balance being commercial. In comparison, kerbside recycling loads in the city totalled 137,400 tonnes (Middleton 2011).

The refuse/recycling operators were accompanied in their collection vehicles. Refuse transfer stations and a recycling plant were also visited to witness unloading. Semi-structured questions were posed, roles were observed, and notes and sketches of road layouts were made. After four operators were observed and interviewed, an executive summary was shared with depot managers to ask follow-up questions. There was a delay of a few months as Company Y wound down, out of concern for possible stress to contractors from the organizational changes. Once the new arrangement was adjudged to be 'bedded in', a fifth operator was interviewed for triangulation.

Participants and their learning of mathematics at school and at work

This section describes findings regarding the participants' learning of mathematics. The experienced orchard manager, Gary, said that he avoided mathematical situations and felt inadequate. 'I just had this thing about numbers; I backed off. People would ask me a simple sum and I'd just go blank'. However, once he moved into management, Gary felt he 'was slipping behind' and in-task calculations would emerge, pressing him for decisions. His partner helped him gradually improve his number sense, and he described how it suddenly fell into place:

> Due to my phobic ways in the past I broke through that ... some strange bloody thing happened because one day all of a sudden something in my brain must have clicked and I looked at something and I thought, oh I've come out with the answer in my head. I was proud, which is crazy for a grown man?

Dave described his school mathematics learning bluntly: 'No, I was crap'. But he then recounted how he had managed 'a multi-million dollar store' involving retail budgets, staffing and stocktaking, '[...] and you're doing maths all the time'. Becoming familiar with orchard work, it seems each manager's quantitative confidence increased with the decisions they were making. When hired, staff are not assessed for numeracy or literacy,, and Gary added that he would not have been employed if tests had been in place.

In the second study, three of the five operators believed they were competent in mathematics. Josh said he had a grasp of the basics of number and measurement, but when maths became more abstract, 'I sort of struggled a bit'. Dennis said that he was comfortable with his learning of school mathematics, but a teaching change impacted on his motivation and subsequent disconnection:

> I thought I was good at maths, liked competition, [and] was fast at working stuff out. This teacher had five of us doing more advanced stuff than the other kids, then [the school] stopped teaching us and we lost interest ... the five of us were no longer challenged.

Brian said that he had 'no problems with maths' at school or afterwards, explaining he had managed a liquor store and employed several staff. Noel said that 'maths is very important in life'; as a contractor, his accountant now 'took care of the maths'. Sean stated that school was insignificant to him and that he had 'learned from life experience, not from education'. The operators did not see the relevance of mathematics in their work. Four had previously driven large vehicles and believed their accumulation of experiences such as increased spatial awareness, estimating loads, and learning street layouts was due to common sense (Coben 2000).

To many people, mathematics appears to have minimal involvement in their practices. Noss, Hoyles and Pozzi (2000, 33) account for this missing mathematical detail as a user's proximity to their own situation, having 'intimate connections with the specificities of their practice'. Although most of the participants agreed that mathematics is important, it did not translate to *what they perceived was relevant* in their own activities. Mathematics is embedded in their work practices and is only drawn on *when required*. Niss (1994, 371) describes this as the 'relevance paradox' where people miss seeing the mathematics they need, and yet 'the social significance of this mathematics [occurs in] ever increasing scope and density'. Noss et al. (2000) suggest that mathematics usage is unpredictable, depending on whether the work is routine or there is change or irregularities. The mathematics only then demands closer attention.

Estimation

The orchard managers and the operators provided ample evidence of estimation in their respective workplaces, while they were drawing on number sense, measurement and spatial reasoning.

Orchard managers and estimation

In this section, practices are described as the seasons unfold in an orchard. Managers' roles include *budgeting, administration, maintenance* and *health and safety* practices. Budgeting covers *staff training, vine pruning, counting buds and flowers, controlling pests*, and *sampling and harvesting*. Some roles require expert staff while there are everyday practices that everyone is expected to competently and safely complete. The first season to be investigated is winter.

Winter pruning

After kiwifruit harvesting ends in autumn, female vines are pruned; male vines are pruned after flowering in late spring. Managers train work teams to carry out orchard tasks such as pruning and thinning. Staff must count accurately and be spatially aware under the vine canopy in order to evenly shape the residual canes. Poorly trained staff generate costs downstream.

The managers' instructions on pruning vary. Dave illustrates desired lengths by using lopper handles while Gary spreads his hands out to illustrate widths. Dave sets a 30cm to 40cm spacing between the canes, while Gary has a 320mm spacing. Little appears written down for staff, although Gary draws diagrams on white boards. Dave's mentor, Rod, has a training zone ('pet bay') and teams train there for orchard practices. Each manager escorts his teams into orchard blocks, monitoring and offering feedback. Dave explained, 'We always mark out the last part they pruned, then go down another row, circle around again and follow up, checking ...'. Gary showed which canes he wanted to protect, leaving 'the first one, or chop the second, leave the next one, chop the next one – get a fishbone effect ... not too crowded'.

After pruning, canes are tied down evenly, so manageable canopies can form. To leave the best buds, Gary described the pre-flowering thinning activity:

> get rid of all the inherent disfigured stuff, flats, fans, short spurs ... could grow to having eight fruit on them (so) bring them down to say three ... you take off the rubbish, knock off the first two and the last one as it's always going to flower later.

It is critical to 'set bud numbers right pre-flower since in spring we need to have our bees visiting each flower many times'. Gary thins early so 'when the flowers pop out, only those I want to pollinate are there'. Staff teams of four walk through each bay, moving shoulder-to-shoulder. They also look for and kill *scale*, a small crawling insect pest that after hatching can crawl a metre in two to five days (Blank, Olson and Lo 1990) before feeding on fluids from each vine.

High or low bud counts per square metre spur a team into action, and managers recognize 'where pruners have been a bit overzealous and they've tidied too much wood'. Gary recalled when a bud count decreased from 38 to 23 buds/square metre, well below the agreed rate of buds/m² between owner and manager. A higher rate was needed to compensate from adjacent bays. Gary sets benchmarks of 15 minutes per bay, although with several staff this reduces to 'about four minutes, then they've got to go to the next bay'. He recalled when one grower had several bays each with 2500 fruit instead of 1200. The excess was removed but the agreed budget for thinning was exceeded. Gary applies a cost per area rate (figure withheld), but he builds in a small margin for any shortfall, so as 'not have to go to the owner and look for an extra two grand to do a tidy up'.

Spring

Having sufficient numbers of buds bursting into flower is critical. When the air temperature reaches 15°C consistently, plants start growing and pest monitoring is essential. With the risk of occasional spring frosts, managers monitor air and soil temperatures to act before frost strikes. Dave's orchard was located at a lower altitude and frosts were rarer. He explained that 'your fruit's going to come off the vine sort of quicker ... [so] the money flow is quicker'.

By thinning vigorous shooting canes, light and air penetrate the canopy for bees to access flowers. As kiwifruit flowers are nectar poor, they do not appeal to insect pollinators (NZKGI 2016), so beekeepers feed bees a syrup mix to keep them foraging among flowers during the day. It is vital to have enough bees at pollination time and Gary explained they used eight to twelve hives per hectare with thousands of bees in each hive. There needed to be almost 100 per cent 'flower burst' before bees were introduced, otherwise as Dave stated 'neighbours' orchards become the beneficiaries of their work'.

When a kiwifruit is cut open, pollination is successful if a size-33 fruit has 'about 1600 seeds in it … drop[s] out nice and oval' (Gary). Size-33 corresponds to 33 similar kiwifruit fitting into a 3.6kg capacity tray (NZKGI 2016). Gary also described flower ratios where some buds produce more than one flower, and a ratio of 1.5 flowers per bud is not uncommon. If the flower (or later fruit) count is too low, then *girdling* may be tried. The circumference of the female kiwifruit trunk is shallow grooved by a cutting chain (EP Prunings 2013, 3) to 'shock [the vine] into producing more buds or flowers'.

Summer

Typical kiwifruit bays (*monitor bays*) are revisited each year and provide estimates of average bud counts. Maps give locations (orchard, block, bay) and expected quantities such as winter bud counts. A crop estimate is used to confirm picking, storage and packaging budgets. With ten monitor bays per hectare spread through a block, some holdings have several hundred monitor bays. Counts also guide urea fertilizer applications in the orchard. A rate of 100kg per hectare (Gary) equates to a handful of urea (or about 40 grams) per vine.

Gary's counter visits monitor bays at least three times a year. If a fourth visit is required she will count one in three bays. The monitor bays are laid out in a 'W' formation.

> Every [monitor bay] is marked – easily visible with painted posts. On the map it'll say West 7, South 4, so … it's seven from the west side and four from the southern side, so you start from there. Then continue [systematically] … and follow the map (Gary).

From this data, managers know whether they will reach the desired numbers of trays per hectare. To estimate final fruit, every monitor bay is counted so managers can see how accurate their figures are. This gives an average buds/m^2 rate, with a higher rate set for the more lucrative kiwifruit gold variety.

Autumn harvesting

Every orchard has records of daily temperatures, picking times, frosts, and yields. Dave refers to records with expected weight ranges for fruit through the season.

> About four weeks before [harvest] … we get seven of the most average-sized fruit and put them on the scales … at that time of the year fruit should be say 73 grams, it might reach that, … but if not, then most them

aren't, they've got to come off … [so] you're putting what that fruit would have got into another fruit.

The brix level (sugar content in solution) determines when fruit is ripe enough to pick. Levels must be above 6.5 per cent before green kiwifruit is picked and 8 per cent for the gold variety (NZKGI 2016). For a sample of 30 fruit, Dave targeted specific vines across the orchard. Each fruit is tested by 'an independent company' to measure brix levels (NZKGI 2016, 80). Another key check is the dry matter (DM) ratio which compares the mass of dry matter (water removed) in a piece of fruit to the total mass of the fresh fruit. In most varieties the DM ratio must be at least 15 per cent before harvest (NZKGI 2016).

A large labour force is required to pick the ripening fruit. The Restricted Seasonal Employees (RSE) scheme invites labour from Pacific nations to harvest. Gary uses a distributed leadership model (Eraut 2007) where he invites a senior bilingual team member to instruct the workers on his behalf. Remuneration for all, from owner to pickers, is tied to harvesting rates; in a good year the volume could reach a median of 10,000 to 12,500 trays per hectare of fruit. Gary described a previous day's picking where one team harvested a six hectare site, filling 889 bins (holding 70 trays or 15 bushels of fruit). With more than 62,000 trays, the 10,000 trays per hectare target was achieved easily.

Gary explained how the picking rate varied with the fruit. 'We want our guys to pick 0.9 of a bin per person per hour of a 15 bushel bin on Gold; [but] they can go faster on green as it doesn't have the same damage issues'. Quality controllers monitor actions of picking teams for damage. With 320 fruit on the top layer of a bin, if they 'find a couple of defects it's not a major issue because the percentage is so low'. However, Gary recounted how yesterday six defective fruit were found on top; 'that's about two percent … you'd go uh oh, then check all the bins around you'. Picking is weather dependent, and if rain is likely, teams harvest into the night to reach their targets.

The orchard managers' estimation practices were seen in counting, in sampling, spatially, and proportionally (rates). The growing and harvesting of exportable kiwifruit, while concomitantly heeding seasonal and environmental factors, illustrates the complexity of their workplaces, and how estimation enables key management decisions to be made.

Refuse and recyclables operators and estimation

Workplace practices of refuse/recyclables operators include *driving*, *collecting*, *emptying* and *safety*. *Maintenance* and *administration* practices are important but less frequent. Since refuse bins (120 litres) are collected weekly, refuse trucks load more bins with heavier loads. Recyclables bins (240 litres) hold commingled paper, glass and some plastics, have generally lighter loads, and are collected fortnightly. A single operator in either role has displaced three or more collecting jobs Some suburbs still use refuse bags, so other companies (not in the study) employ runners and a driver.

The technology

Each vehicle has standard gear sticks, pedals, mirrors, and steering wheels, on each side of the cab, enabling dual-control. A joystick engages the drive chain for the hydraulically operated arms (see Figure 2.1) to reach, grab, lift, empty and return bins to the berm (roadside verge). As material drops into the hopper, it is moved to the rear by a hydraulic paddle. The weight of each load is measured by scales in the hopper that are read in the cab. Three other on-screen figures are displayed. A counter tallies the number of bins emptied; the screen in Figure 2.3 shows that 139 bins have been emptied. The current time and the hours in use since the last hydraulic service are also shown. Radio telephones and GPS systems enable contact and locating by other operators and depot staff respectively.

Compliance practices for recyclers

Refuse collectors' loads are full when no more refuse may fit. Recycling operators have load limits however, faced with a compliance weight arrangement with the city council, the collection company and the recycling depot. Each truck is weighed as it enters and exits the recycling depot. If the load (see Figure 2.2) is above the compliance weight, then it is notified as being over-compacted (above a commercially sensitive rate in kg/m^3) that theoretically takes recycling depot staff more time to separate. Financial penalties are imposed if overweight loads enter the depot. Company X pays for its employees but the thirteen contracted recyclables operators must meet any costs themselves.

FIGURE 2.1 The truck's side chain extends to berm so hydraulic arms can lift a recycling bin, as seen from rear of cab (source: P. Kane)

FIGURE 2.2 The hopper of a recyclables truck emptied of commingled materials (source: K. Chana)

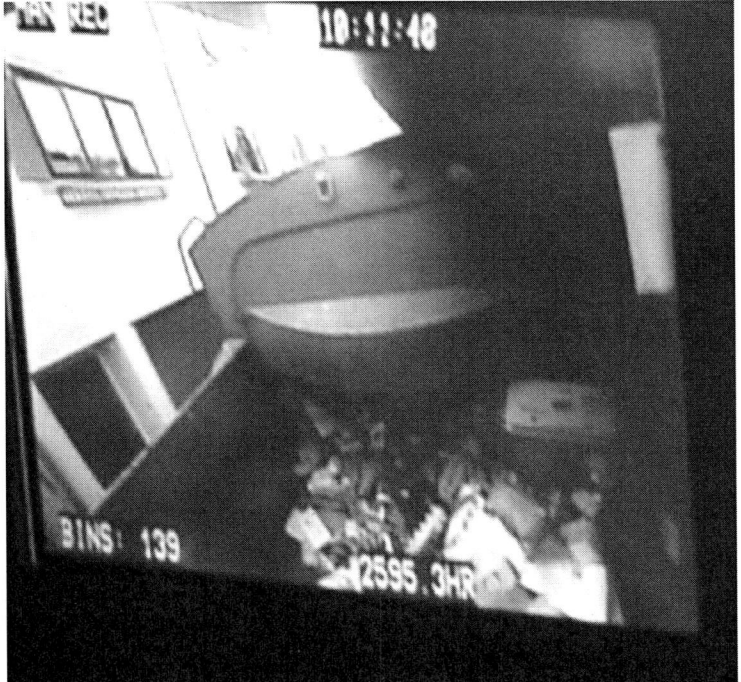

FIGURE 2.3 Truck cab monitor with bin counter and clock (source: K. Chana, taken while vehicle in motion)

Bin numbers, time and seasonal practices

Being able to estimate the number of bins is important. Operators' estimates usually rely on neighbourhood familiarities. Sometimes they subitize (by recognizing up to four items without needing to count), then combine clusters of three to four bins. The count is challenging if there are many dwellings down a driveway, although Dennis matches bin numbers to the number of letter-boxes. Operators choose the street order – this day Dennis first completed what he called 'a tricky street' since he found the refuse bins there are always heavier. He expected the bins in adjacent streets were lighter, so more bins are taken in his next load. When the end of a load looms, territorial familiarity is vital. Sean explained, 'There are a couple of streets I don't go into when you're almost full … just too many bins", so those streets could 'wait until the next load'.

Bin numbers are also connected with time and timing. Operators begin at 5:30am but cannot collect residential bins before 7am, so they start along commercial areas and main roads. Routes include estimates of when they arrive, and they try to avoid rush hour traffic. As Josh explained, starting times impact, so 'before 7am you get a good flow, but if between 7 to 8 o'clock, then add 15 to 20 minutes'. Seasons also influence load sizes with warmer months having more loads, since there is more outdoor entertaining by residents. Josh provided records from May where one run had a four loads total of 965 bins. The same run in July however, had 841 bins, so, with there being 120–130 fewer bins, Josh planned for just three loads of 270–280 bins each. Public holidays mean long weekends, so collections are delayed by a day. Saturdays become collection days, and Noel noted that although loads are bigger, traffic volume is usually less, so 'I'm ten minutes ahead of myself … I usually get to [this street] later, but today we are here at 11am'.

Since refuse operators empty more bins, Dennis believes they have steeper learning curves than recyclers, but also have 'a faster rate of improvement'. Experienced operators like Josh use the rates to predict, such as 'ten minutes collecting here, fifteen to twenty minutes there', and he was observed to be accurate to within two minutes each way. Dennis said he could 'empty 1000 bins before lunchtime', although in busier times, he noted if refuse loads have 10 per cent more bins, that adds an extra thirty minutes to his run. With the arms on refuse trucks lifting and dropping every twelve to fifteen seconds or about 1400 times a day (cf. recyclers ~1000 times per day), predictably refuse truck belts, tyres and brakes wear faster.

Practices involving locations

Bin locations often provide challenges. Operators explained there were socioeconomic aspects associated with bin contents and that this affects (un) loading. Less wealthy areas often have relatively heavy loads if householders fill recycling bins with refuse, thereby contaminating a load. Inner-city bins take twice as long to collect as suburban bins owing to parked cars along narrow roads or trees on berms blocking lifts. Josh described this as 'the spatial awareness thing', although operators are always contending with angles, heights and distances. Schools are avoided at opening and closing times, while road works and construction zones require patience and good sense. In suburbs where refuse bags are still used, recycling and refuse trucks might occupy the same road, runners/bag collectors usually move faster, so an informal spatial courtesy has evolved enabling the faster vehicle to pass.

Knowing territories and expected bin numbers, operators usually know the best routes and estimate their *fuel* needs (especially contractors who purchase their own). Operators will wait until reaching a level road to read scales accurately. When a load is almost full, they plan which streets will have enough bins to complete the load. Josh recounted that 'from experience I knew there were three to four bins in the no-exit road left to complete my second load of 275 … so one more load would finish the run'. But no-exit roads have their own demands. The turning circle of a cul-de-sac may have several driveways off arced berms. Operators usually know the best times to visit these streets, but they must also be vigilant of other transport such as courier vans. Industrial culs-de-sac may have several businesses off each driveway, with their staff parking along berms blocking bins from collection vehicles. Despite the 2.5– 3m reach of the hydraulic arms, they often have little option but to alight from the vehicle to move bins. Alighting from truck cabs is risky as with the 1.7m height of the seats above ground, it is less a case of getting in and out, and rather more of ascending and descending.

Practices around measuring

Recyclables operators monitor their loads when about two-thirds full using the hopper-mounted camera. Originally, the vehicles did not have scales and Josh noted 'in the early days, before our scales, when still learning … some guys were [paying] $600 to $700 a month' in penalties for having overweight loads. Fortunately, this changed once scales were installed, but Josh had become attuned to his estimations. His empty truck weighed 10,070kg, and by including his own weight 'and a full tank of diesel, the full weight will be 10,260kg; if the tank is quarter full, [the total] is a bit under 10,200kg'.

There were times when loads and bin numbers deceive even experienced operators. Josh recalled how on a previous run his three loads had totalled almost 900 bins. The bin numbers were similar but one load was disproportionately heavier than the others.

> I've tried to do three runs of 300, and come in underweight, but gone over. Underweight for the day but overweight for an individual load … one I'll be pinged for. Funny enough, you see the first load has the most [310]

bins, but only weighed 3.78t. The middle load [only 294 bins] weighed 4.12t ... less bins than the first one but still slightly more [weight], ... still over 200 kilos too heavy.

Based on previous trends, Josh anticipated that 300 bins on average would be a compliant load, but he underestimated, and the extra weight left him with an unwelcome cost.

Summary of practices from each case

Participants revealed how estimation was used within their workplace practices during *pruning, counting buds and flowers*, and *sampling and harvesting* for orchard managers, and *loading, weighing, locating* and *complying* for collection operators. Table 2.1 summarizes several key instances where estimation was carried out in work practices.

In the workplace practices presented above, estimation nests within the mathematics of *number sense, measuring, data handling* and *spatial awareness*. Alongside the elemental workplace motives, meanings and vigilance for quality, the quantitative and spatial choices made by participants influence each operation. For instance, decisions of orchard management have been found to make larger impacts on orchard gate returns than where an orchard is located (Woodward 2014).

Discussion

Visibility

Not only is mathematics almost invisible to many people (Niss 1994, Coben 2006), but there are also instances which people do not perceive are 'legitimately' mathematical (Cockcroft 1982; Coben 2000). Estimating may not be considered as *doing mathematics* if a problem is not solved *by using a traditional formula or an equation*. Perhaps individuals *choose not to apply* 'a mathematical gaze' (Lerman 2000, 21) in their situations, and this could account in part for numeracy invisibility. Josh's exit strategy is an example:

> Usually I'd go in here, do a figure of eight, so I'd go left – left – left – left, back to this point, then straight across, then left – left – left and come back out of that street again. That'll put me over the number of bins I want to have for my 300, ..., so I could do that but it means I'll return on the third run and start half a street down. I'm a Libra [laughs], I like things being equal, so I'll try get as close to 300 doing it this way and start afresh on a completely different area for my third run. There's no maths involved in that at all.

TABLE 2.1   Summary of key instances of estimation in each workplace

Orchard managers	Refuse/recyclables operators
Use of artefacts such as lopper handles and hand spans to indicate lengths and widths	Subitize then add bin bunches (i.e. several bin lots combine to 11,12 or 15,16 bins).
Fishbone effect used to model growing canes by pruning alternate canes.	Following public holidays, a delayed Saturday bin collection has almost 100 per cent turn out
Pre-flower thinning to leave the best buds; e.g. an eight-bud cane was pruned to three or four buds.	An extra 10 per cent more bins might equate to an extra 30 min collecting time
Between 8–12 beehives per hectare and expect bees to visit flowers up to 40 times.	Suburban collecting is about twice as fast as inner-city collecting
Use of monitor bays to historically represent typical bays across the orchard, evenly spaced across the blocks.	Contractors refuel when it is convenient whereas employed operators refuel on their return to the depot each evening
Pest control of junior scale between 500mm to a metre across the wood in the canopy	Time taken to get to tip-off or schools and return to territory, avoiding rush hours.
Bud counts per square metre when there are too many so some buds must come off; or when not enough buds so extra buds must be found from elsewhere (to preserve budget)	Recognise approximate numbers of bins which will be on certain roads, and when to collect there or delay for the next load, depends on current load status (weight).
Record spring temperatures and thus remain vigilant for frosts, with preventative equipment on standby	Recyclers vigilant when they get to about 70 per cent of their load (or near the 3t weight on their scales) – need to keep weighing
Continue thinning and enable access for bees to visit flowers (many times).	Recognise 10 min here and there and approximate bin numbers expected in roads
Bud, flower, and fruit counts used to estimate crop and budgets for picking, packing, storage and transport.	First load often indicates the way a run will unfold in bin numbers and weights; however operators still monitor scales.
Desired fruit weight ranges in weeks leading up to picking, directs which fruit will not make the grade and will be removed, leaving optimum fruit to mature.	Time lost when having to alight to shift bins, but also so as not to block other drivers (e.g. couriers), and clean up spills – up to 10 min taken out of a run.
Brix percentages and dry matter ratio figured from representative samples before picking.	Experienced operators can usually sense if a bin is overweight if the joystick strains more.
Known bin capacities (70 trays per CA bin) can be used to approximate the trays per hectare to match budgeted volumes.	Planning when considering route(s) particularly for exit strategy to tip off, so fuel, time and mileage are saved.
Quality-control estimates proportion of damaged fruit on top layer of bin.	Heavier weights in bins and loads over from spring to autumn with many more outdoor events and functions.

The study participants already have many practices to contend with, including financial and safety implications, quality outcomes, and eliminating on-site errors, so looking through a numeracy lens is not their priority.

Numeracy as situated social practices

Each case provides evidence for numeracy as a model of socially situated practices. Participants' estimations and decisions are always made in situ, influenced by seasonal cycles, local traffic and authority regulations. There is evidence of an ideological model of social practices through commercial and civic objectives, with each workplace regulated by compliance, authority, safety and responsibility. Drawing on experiences, records and trends in either organization, their practices are historically, environmentally and culturally situated.

The NLS model is however, not as tidy a fit for numeracy as it is for literacy/literacies. There are strong connections between mathematical strands, and their coexistence needs neither reduction (Tomlin 2006) nor camouflage. Instead, a broad dynamic flexi-model of numeracy practices that bends, twists and responds to each relevant situation is perhaps apt. Individuals select numeracy practices such as estimating to weave into their intentions and "different trajectories" (Gal 1999, 27). Not everyone (e.g. Gary), however, is confident to draw on mathematical ideas. Many adults have self-doubts, and it takes time and considerable effort to learn mathematical nuances, precision, abstractions and notations, and *transfer* these to other contexts. Coben (2006, 101) notes that

> a significant difference between numeracy and literacy (is) the fact that reading and writing a language one already speaks is not analogous to 'doing mathematics' or 'being numerate', because most people do not 'speak mathematics' as a first language; they need to be taught it.

Hoyles et al. (2002, 9) caution that workplace mathematics has the immediate goals of 'remaining competitive' and 'maintaining operations'. From these cases, decisive steps would be taken if too many buds were thinned over several kiwifruit bays, or if a spillage occurred, since timely and costly outcomes await (Cockcroft 1982). Each participant has built up niche proficiencies (see Table 2.1), so when an unfamiliar situation arises, they may confront the issue by delving into a cache of strategies. Individuals may still be challenged, however, to articulate exactly how and what they do when drawing on those established tacitly known (Eraut 2004) practices, such as estimation. As refuse operator, Dennis, observed:

> You know, I think a lot of things you do you take for granted – you know you do it, just second nature really. Just estimate all the time. There are things some people do in their work but it [estimating] helps.

Conclusions

Although a model of socially situated practices is perhaps a better fit for literacy than for numeracy, it still provides a useful form when understanding that numeracy practices are always purposeful and occur in tandem with workplace practices. There is plenty of evidence of historically, environmentally and culturally situated practices with each participant relying on previous records, maps and artefacts. The numerate actions of the orchard managers and the collection operators are always situational, and are part of a larger agenda. Participants respond to social and hierarchical demands and deal concomitantly with the realities of commercial and civic interests, and on-site or distant authorities who have requirements for compliance, quality and health and safety. Each participant goes beyond basic calculation skills or measurement; they have a grasp of location and space, a reliance on a reasonable sense of size and direction, and an ability to make regular on-the-spot estimates. A possible departure from an explicit social practices model such as NLS occurs when participants cannot detect mathematics such as estimation in their work. In fact, the making of estimates by the participants is so common that it is almost not counted as being *real* mathematics, even though it is clearly a business-as-usual practice for them.

Acknowledgements

I wish to thank the orchard managers, the operators and their site managers for their time and the information they provided about their respective industries. I acknowledge the support of Shelley Rose at Bay of Plenty Polytechnic, and Associate Professor Pat Strauss, Dr Lynn Grant, Kevin Roach and the late Frank Smedley at AUT.

References

Adams, T.L. and Harrell, G. (2010) "A study of estimation by professionals at work" *Journal of Mathematics and Culture*, 5(2): 1–15.

Baker, D. (1998) "Numeracy as social practice" *Literacy and Numeracy Studies*, 8(1): 37–50.

Barton, D. (2006) "Significance of a social practices view of language, literacy and numeracy". In Tett, L., Hamilton, M. and Hillier, Y. (eds), *Adult Literacy, Numeracy, and Language* Maidenhead: Open University Press.

Blank, R.H., Olson, M.H. and Lo, P.L. (1990) "Armoured scale (hemiptera: diaspididae): aerial invasion into kiwifruit orchards from adjacent host plants" *New Zealand Journal of Crop and Horticultural Sciences*, 18(2): 81–87.

Campbell, H. and Haggerty, J. (2008) "Kiwifruit". *Te Ara – the Encyclopaedia of New Zealand* Retrieved 8 February 2009 from http://www.teara.govt.nz/en/kiwifruit.

Chana, K. and Kane, P. (2010) "Situated numeracy in everyday contexts" Presentation at the Australian Council for Adult Literacy 33rd Annual Conference: Hands up... Hands on... Darwin, Australia: ACAL, September 10–11.

Chase, W.G. (1983) "Spatial representations of taxi drivers". In Rogers, D. and Sloboda, J.A. (eds), *The Acquisition of Symbolic Skills* New York: Plenum Press.

Coben, D. (2000) "Mathematics or common sense? Researching 'invisible' mathematics". In FitzSimons, G. (ed), *Perspectives on Adults learning Mathematics: Research and Practice* Dordrect: Kluwer Academic Publishers.

Coben, D. (2006) "Socio-cultural approach to adult numeracy: Issues for policy and practice". In Tett, L., Hamilton, M. and Hillier, Y. (eds), *Adult Literacy, Numeracy, and Language*, Maidenhead: Open University Press.

Cockcroft, W.H. (1982) *Mathematics Counts: Report of the Committee of Inquiry into the Teaching of Mathematics in Schools* London: Her Majesty's Stationery Office.

EP Frunings (2013) *EastPack* Update 127. Retrieved 18 December 2016 from http://www.eastpack.co.nz/vdb/document/23527.

Eraut, M. (2004) "Informal learning in the workplace" *Continuing Studies in the Workplace*, 26(2): 247–273.

Eraut, M. (2007) "Learning from other people in the workplace" *Oxford Review of Education*, 33(4): 403–422.

FitzSimons, G.E. (2000) "Mathematics and the vocational education and training system". In Coben, D., O'Donoghue, J., and FitzSimons, G.E. *Perspectives on Adults learning Mathematics: Research and Practice,* Dordrecht: Kluwer Academic Publishers.

Gal, I. (1999) "Numeracy education and empowerment: Research challenges". In van Groenestijn, M. and Coben, D. (eds), *Adults Learning Mathematics 5. Mathematics as part of lifelong learning: Proceedings of the fifth International Conference of Adults Learning Mathematics* London: Goldsmiths College, University of London and ALM.

Gee, J.P. (1996) *Social linguistics and literacies: Ideology in Discourses* (2nd edition), London: Taylor & Francis.

Gee, J.P. (2010) "A situated sociocultural approach to literacy and technology". In Baker, E.A. (ed), *The New Literacies: Multiple Perspectives in Research and Practice* New York: Guilford Press, New York.

Hennink, M., Hutter, I. and Bailey, A. (2011) *Qualitative Research Methods* London: SAGE Publications.

Hodgen, J. and Marks, R. (2013) *The Employment Equation: Why Our Young People Need more Maths for Today's Jobs,* London: The Sutton Trust. Retrieved 18 January 2018 from https://www.bl.uk/britishlibrary/~/media/bl/global/social-welfare/pdfs/non-secure/e/m/p/employment-equation-why-our-young-people-need-more-maths-for-todays-jobs.pdf

Hogan, J. and Morony, W. (2000) "Classroom teachers doing mathematics in the workplace". In Bessot, A. and Ridgeway, J. (eds), *Education for Mathematics in the Workplace* Dordrecht: Kluwer Academic Publishers.

HortResearch Rangahau Ahumara (The Horticulture & Food Research Institute of New Zealand) (2007) *Fresh Facts 2007: New Zealand Horticulture*. Retrieved 18 January 2018 from http://www.freshfacts.co.nz/files/fresh-facts-2007.pdf.

Hoyles, C., Wolf, A., Molyneux-Hodgson, S. and Kent, P. (2002) *Mathematical Skills in the Workplace: Final Report to the Science, Technology, & Mathematics Council,* London: Institute of Education, University of London.

Lave, J. (1993) "The practice of learning". In Chaiklin, S. and Lave, J. (eds), *Understanding Practice: Perspectives on Activity and Context* New York: Cambridge University Press.

Lave, J. (1996) "Teaching, as learning, in practice" *Mind, Culture and Activity*, 3(3): 149–64

LeFevre, J., Greenham, S. L. and Waheed, N. (1993) "The development of procedural and conceptual knowledge in computational estimation" *Cognition and Instruction*, 11(2): 95–132.

Lerman, S. (2000) "The social turn in mathematics education research". In Boaler, J. (ed.), *Multiple Perspectives on Mathematics Teaching and Learning* Westport, CT: Ablex.

Mainland, C.M. and Fisk, C. (2006) "Kiwifruit" Horticulture Information Leaflet 208, North Carolina Cooperative Extension Service. Retrieved 18 January 2018 from https://content.ces.ncsu.edu/kiwifruit

Merriam, S.B. (2009) *Qualitative Research: A Guide to Design and Implementation* San Francisco, CA: Jossey Bass.

Middleton, B. (2011) *Auckland Council Waste Assessment: Data update*. Prepared for Auckland Council Solid Waste Business Unit. Auckland: Waste Not Consulting.

Niss, M. (1994) "Mathematics in society". In Biehler, R., Scholz, R.W., Sträßer, R. and Winkelmann, B. (eds), *Didactics of Mathematics as a Scientific Discipline* Dordrecht: Kluwer Academic Publishers.

Noss, R., Hoyles, C. and Pozzi, S. (2000) "Working knowledge: Mathematics in use". In Bessot, A. and Ridgway, J. (eds), *Education for Mathematics in the Workplace* Dordrecht: Kluwer Academic Publishers.

NZKGI (New Zealand Kiwifruit Growers Incorporated) (2016) *The 2016 Kiwifruit Book: A Resource for New Zealand Secondary School Teachers and Growers New to the Kiwifruit Industry*. Mount Maunganui: NZKGI

Siegler, R.S. and Booth, J.L. (2005) "Development of numerical estimation: A review". In Campbell, J.L.D. (ed.), *Handbook of Mathematical Cognition* New York: Psychology Press.

Smith, J.P. (1999) "Tracking the mathematics of automobile production: Are schools failing to prepare students for work?" *American Educational Research Journal*, 36(4): 835–878

Sowder, J. (1992) "Estimation and number sense" In Grouws, D.A. (ed.), *Handbook of Research in Mathematics Education, Teaching, and Learning* New York: Macmillan.

Stein, D. (1998) "Situated learning in adult education" *ERIC Digest 1998–3*. Retrieved 14 February 2013 from http://www.ericdigests.org/1998-3/adult-education.html

Street, B.V. (1995) *Social Literacies: Critical Approaches to Literacy in Development, Ethnography and Education* London: Longman Group.

Street, B.V. (2012) "New literacy studies". In Grenfell, M., Bloome, D., Hardy, C., Pahl, K., Rowsell, J. and Street, B.V. (eds), *Language, Ethnography and Education: Bridging New Literacy Studies and Bourdieu* New York: Routledge.

Tomlin, A. (2006) "If you can make it you can own it" *RaPAL Journal (Research and Practice in Adult Literacy)*, 59: 31–37.

Triantafillou, C. and Potari, D. (2010) "Mathematical practices in a technological workplace: the role of tools" *Educational Studies in Mathematics*, 74: 275–294.

Woodward, T.J. (2014) "How important is location in determining kiwifruit orchard returns: Spatial variation in 'Hayward' kiwifruit orchard gross revenue within a growing region across seasons" Kellogg Rural Leaders Programme 2014 and Lincoln University, Canterbury, NZ. Retrieved 18 January 2018 from https://www.kellogg.org.nz/uploads/media/Woodward-Tim-How-important-is-Kiwifruit-orchard-location.pdf.

Yin, R.K. (2009) *Case Study Research: Design and Methods* (4th edition) Thousand Oaks, CA: SAGE Publications.

Zevenbergen, R. and Zevenbergen, A. (2004) "Numeracy practices of young workers". In Holnes, M.J. and Fuglestad, A.B. (eds), *Proceedings of the 28th Conference of the International Group for the Psychology of Mathematics Education*, Bergen: Bergen University College. Retrieved 22 July 2009 from www.emis.ams.org/proceedings/PME28/RR/RR093_Zevenbergen.pdf.

3

BUILDING STONE WALLS

A case study from the Philippines

Wilfredo Vidal Alangui

Introduction

The Cordillera region of northern Philippines is home to diverse indigenous peoples with distinct cultural practices, languages and knowledge systems. One thing they have in common is that, for many centuries, the Cordillera peoples have subsisted on a livelihood based on rice cultivation, by carving extensive rice terraces out of the slopes of mountains and rugged terrain, a proof of the capacity of human beings to adapt to the limitations of their environment.

Rice terraces are not unique to the Philippines, they are also found in China, Indonesia and Korea. However, the Cordillera rice terraces have been described as 'the product of agricultural endeavors that may be classed among the most intensive and efficient in the world' (Bodner 1986, 1). Owing to the steep terrain in the region, one important aspect of rice terracing in the Cordillera is the building of stone walls that hold the rice paddies. It is this practice of building stone walls that is the focus of my investigation of numeracy as a social practice.

Mathematics and terracing

This chapter revisits my earlier work (Alangui 2010) where I drew parallelisms between the practice of building stone walls and mathematical practices, searching for 'alternative mathematical concepts' in the practice. Barton (1996) argues that the category 'mathematics' is not common to all cultures, making it problematic to talk about the 'mathematics' in a specific practice or context if the practitioners belong to a culture that does not have this category. Instead, he uses 'QRS systems', which he sees as a systematic approach to understanding and communicating aspects of our 'quantitative, relational and spatial' realities.

I have stressed 'the need to locate the meaning of the quantitative, relational and spatial ideas in their cultural context' (Alangui 2010, 68). This time, I will argue that building stone walls is an example of how numeracy can be viewed as a social practice.

Because mathematical knowledge is generally considered highly developed, I have argued (Alangui 2010) that a good place to investigate mathematical knowledge within a specific culture will be among cultural practices which are similarly highly developed and systematised. In the Cordillera, building stone walls is a highly systematised practice, developed over hundreds of years, which grew and continues to grow with innovation and creativity.

Researching terracing ethnographically

This research was conducted in 2002 with update visits between 2005 and 2011. Baker (2008) and Adam, Alangui, and Barton (2010) contend that numeracy as a social practice necessitates the use of ethnography as a research method that is socially grounded and humanistic in the study of the interplay of cultural practice and mathematics. Thus ethnography was used to document the indigenous knowledge of rice terracing and the associated practice of building stone walls. Qualitative methods for collecting data were employed, including interviews, oral history and participant observation, complemented by secondary research. Ilocano, my first language and considered the *lingua franca* in the region, was used in interviews and discussions. A research aide who spoke the local language Kankana-ey accompanied me during fieldwork in case some discussions needed translating.

There were 28 main respondents, twenty (20) males and eight (8) females, classified as *stone wallers,* acknowledged expert practitioners, interviewed about their knowledge of stone walling; *female and male farmers,* interviewed about their knowledge of rice terracing and stone walling; and *professionals* whose main source of livelihood is not farming, interviewed about their knowledge of rice terracing. This classification is not mutually exclusive; in these communities, it is common for a person to assume a multiplicity of roles – a teacher could be farmer and also skilled stone waller.

My presence in the community allowed me to assume different roles – complete observer, watching some farmers clean a stone wall, direct water to a rice paddy or plant rice; complete participant, weeding and watering a vegetable garden; and participant observer, trying out the repair of a stone wall. I was a participant in the social and cultural world that I was observing. And I knew that I was also being observed.

Context of the research

One of my assumptions is that cultural systems evolve out of the needs of different peoples in different environments. As shown in my work in two villages, Agawa and Gueday, the Cordillera peoples' practice of rice terracing came about and persisted as they responded to a combination of economic, sociocultural, historical and environmental factors (Alangui 2010).

The region and the peoples

Considered as the most extensive system of highlands in the Philippines (Scott 1974), the Cordillera region contains many peaks exceeding 2,000 metres above sea level, with rolling hills and stretches of river valleys along its flanks (Cordillera Peoples' Alliance 2003). It contains the headwaters of most of northern Luzon's numerous big rivers. In the whole of the region, agricultural terraces cover an estimated 28,000 hectares (Spencer 1952).

The Cordillera region is home to around 1.72 million indigenous, non-indigenous and migrant peoples, based on the 2015 Philippine population census. Though diverse, the indigenous peoples of the Cordillera have many things in common. They live in a defined territory, now known as the Cordillera Administrative Region, the geophysical features of which are similar in most parts of the region. They have gone through, and are going through, the same historical experiences. They have shared, and continue to share, the same social, cultural, economic and political concerns. Most of the indigenous peoples identify themselves primarily with the *ili*, a self-identifying community with a specific territory – the ancestral land. While there are diverse types, an *ili* usually consists of a closely-knit cluster of villages, or a core village and its outlying hamlets, within a more or less defined territory (Cordillera Peoples' Alliance 2003).

The people of Agawa and Gueday belong to the Kankanay group, one of several indigenous groups in the region collectively referred to as *Igorot*, a word that comes from the root word *'golot'* meaning 'mountain chain' (Scott 1966). The prefix *'i'* means 'people of' or 'dwellers in'; thus, *Igorot* means 'people of the mountains', 'mountain dwellers'. During the Spanish expeditions to the region in search of gold, the term was already in common use (Scott 1966). While there is danger in lumping together the different peoples of the region, the use of the word suggests both a common identity and history. Strategically, it has helped galvanise the peoples' political struggle for self-determination.

Nevertheless, the use of the word *Igorot* continues to elicit discomfort for some people because of how it has been used to constitute a demeaning image of the region's inhabitants, that for many years positioned the *Igorot* as second-class citizens. The history of the Cordillera is a narrative of 'conquest and resistance, exploration and occupation, trade and proselytism, conflict and revolt, as well as of cultural persistence and change' (Conklin 1980). Colonisation under Spain and the United States resulted in disruptions in the lives of their colonial subjects. Past and current misconceptions about the *Igorot* are a product of a lowland–highland dichotomy that was perpetuated and exploited by the Spanish colonisers, and later by the American colonisers, the impact of which was 'subtler, more tragic and longer lasting' (Scott 1974, 6; see also Alangui 2010). The Philippine government, the educational system and the church at various times in the country's history have propagated a derogatory view of the *Igorot*. There are efforts to correct these misrepresentations, but misconceptions and incorrect images and stereotypes about the *Igorot* continue to be portrayed, especially in the media (Bauzon 1999).

The Igorot *worldview*

But the *Igorot* were able to hold on to their unique worldviews, beliefs and knowledge systems, although they did not escape the 'acculturative and assimilative processes they have been subjected to in the course of the Spanish and American colonisation, and the onslaughts of modernisation in the late twentieth century' (Bauzon 1999, 8). Some knowledge was lost or changed as a result of the encounter, but by and large, the general features of the *Igorot* worldview persist to this day.

The *Igorot* of the Cordillera, like many other indigenous peoples, have a worldview that is based on:

> seeing the individual as part of nature; respecting and reviving the wisdom of elders; giving consideration to the living, the dead, and future generations; sharing responsibility, wealth, and resources within the community; and embracing spiritual values, traditions, and practices reflecting connections to a higher order, to the culture, and to the earth.
> *(Sefa Dei, Hall and Rosenberg 2000, 6)*

Such a worldview is rooted in the peoples' relationship with the land; land is inextricably linked with their identity and survival as distinct and diverse peoples. The relationship of the *Igorot* with their environment may be summed up by a belief that every generation has the right to avail itself of the resources offered by the land. Because of this belief, the *Igorot* have developed complex strategies to develop and protect their natural resources, strategies expressed in their customary laws and practices.

The villages of Agawa and Gueday

The research sites Agawa and Gueday are two of the fourteen villages of Besao. Besao is in the western part of Mountain Province whose mountains are more than 1000 metres above sea level (OPSRA 1995). It has a dry season (November to April) and a wet season (May to November), and a cold climate with strong winds usually from November to February (Fiar-od 2001). While signs of modernisation are visible and contact with the outside world has increased dramatically through the years, Besao may be considered as one of the areas in the region where cultural persistence is still strong. It is this context that provides a backdrop to the practice of rice terracing and its associated practice of stone walling in this region.

Both villages belong to what is called *i-Agawa* communities that include three nearby villages. Gueday historically had one of the biggest populations among the villages in Besao. In 2000, Gueday had 1172 (727 female and 445 male), while Agawa had a total of 407 residents (186 female and 221 male).

The cultivation of irrigated rice, grown in extensive rice terraces, provides subsistence for most of the people in both villages. Corn, sugarcane and root crops like peanuts and taro are also grown. In Gueday, vegetable production, handicraft and labour are sources of cash in the community, while in Agawa cash comes from labour, livestock and retail business. There is a small percentage of regular wage earners in both areas. The pine forest and the mossy forest found in the area surrounding the *i-Agawa* communities are central to the survival of the people. They contain hardwood used for timber and farm tools and provide medicinal plants and hunting grounds for several wild animals (Institute of Environmental Science for Social Change 2003). The people are aware of the importance of the forests as a provider of water.

The *i-Agawa*, though Christian, continue to adhere to a traditional religion (Tauli-Corpuz 2001; Scott 1974). Among the key elements of this is the belief that land formations, bodies of water, rocks, etc, host spirits who protect them from pollution or destruction, and that deities have particular roles to play. There is a belief in the Creator (*Kabunyan*); prayers of intercession are done through the ancestral spirits. Ritual pervades the life of the *i-Agawa*: there are rituals for each stage of the agricultural cycle in which specific roles are played by male elders, older women and young men and women. The *amam-a* or male elders who facilitate the various rituals occupy a respected position in the village; a ritualist may also be a highly skilled stone waller, as well as an elected political leader in the village.

The *dap-ay*, which is both a physical structure and an indigenous process of decision-making, continues to be an important socioreligious and political institution. It comes under the direction of a council of elders, whose tasks include coordinating activities of the agricultural cycle, settling disputes within and outside the communities (for example, boundary issues), regulating inter-village affairs, and coordinating rituals for *dap-ay* members (Brett 1985). The *dap-ay* serves as an important institution of learning (Alangui 1997) where the young learn the values of the *i-Agawa*. Gueday has seven *dap-ay*, Agawa has six.

It is in this complex environment that the rice terracing practice and stone walling are situated and maintained. Their importance is seen not only in the various rituals associated with the practice, but also in the numerous legends that pertain to rice, rice cultivation and stone walling. For example, oral history tells of the coming of rice to the community: a goddess who pitied the people of Besao for mainly subsisting on *camote*, or sweet potato, disguised herself as a tired and starving old woman and showed herself to a woman named Calindo who gave her *camote* to eat. Grateful and satisfied, she gave Calindo a basket of *palay* (rice grains) for her kindness, with the instructions, 'Some of it you sow for planting in fields with water, and the rest you pound and cook for food' (Staff of Besao Central School 2002).

The papayeo *of Agawa and Gueday*

In the local Kankana-ey language, a rice paddy is a *payeo* and a cluster of rice paddies *papayeo*. *Papayeo* surround the villages of both Agawa and Gueday. It is highly desirable for every person to own a *payeo*, no matter the size. A *payeo* is generally acquired from parents once a young person gets married and starts a family of her/his own. Having been handed down from one generation to another adds to the value of the *papayeo*: 'our ancestors built the terraces for us' (Alangui 1997; Florendo and Cardenas 2001).

Numerous studies have documented the social and cultural as well as the economic importance of the *papayeo* for the peoples in the Cordillera. Rice terracing is integral to the identity and survival of the *Igorot* as distinct peoples. It continues to be woven through the economic, cultural life, political and social organisation and cognitive systems of the Cordillera peoples (Florendo and Cardenas 2001). It integrates technical and agricultural principles with social and cultural knowledge. The associated practice of stone walling is a vital element of this system. Stone walling is an example of how people can adapt to and manipulate their environment, and in the process develop complex knowledge systems that are vital to their cultural identity and survival.

Much of the appeal of the *papayeo* is due to the impressive panorama they offer to viewers from a distance; *papayeo* of different sizes and shapes dominate the landscape, spread out, as if randomly, straddling mountains and filling up plateaux and valleys. Each season offers a different visual delight, but most striking is when the grains are fully-grown and have turned yellow under the sun. This is why the *papayeo* in the Cordillera continue to attract visitors to the region.

There is no consensus on the origins of rice terracing in the Cordillera; competing theories have been put forward by (mostly foreign) historians (Alangui, 2010). One suggestion (Keesing 1962) is that the Cordillera terracing developed after the arrival of the Spaniards in 1521. Scott (1966) and Conklin (1980) also suggested that rice was not indigenous to the region. Bodner's archaeological work (1986, 6–8) showed that the Cordillera had 'a pre-rice agricultural economy' based on the dry cultivation of grain crops; the two dominant crops at present, rice and sweet potato, are 'relatively late additions to the local crop inventory', possibly co-evolving in the sixteenth century coinciding with the intensification of agriculture in the region. She demonstrated that by ca. AD 600, permanent villages were already established in the area. Along with craft production and manufacturing came 'circumstantial evidence' that agricultural practices such as terracing and fencing with stone walls were already being done (Bodner 1986, 466). If this is correct, it can be surmised that the extensive terracing systems that now exist all over the Cordillera developed gradually, originally utilised for growing taro and other root crops but later adapted for growing wet rice. As rice cultivation became common in a community, more walled terraces were built.

Changes in farming and land resource use in the context of domestic livelihood strategies have taken place (Preston 2003). One recent change is the shift to vegetable production motivated by the need to generate cash for the family (Florendo and Cardenas 2001); however, this is not widespread in Agawa and Gueday. Another change, perhaps more serious, is the increasing shortage of labour needed to work the fields, caused by an exodus of family members normally to urban centres, either for work or for education. As a woman elder, one of the respondents, said: 'Ubbing ket pumanaw. Puro lallakay ken babbaket ti mabati nga agtalon' ('Children are leaving. Only the old men and women are left behind to till the land'). The lack of people to farm has disrupted the agricultural cycle, as one family now takes care of more than one *payeo*, making it difficult to prepare the fields in time for synchronised farming. Another factor causing disruption of the agricultural calendar in the whole region is the introduction of new varieties of rice, which grow faster than the traditional rice, and with these, the increasing use of chemicals and pesticides.

Despite these changes, the cultural tradition of rice terracing persists, and the practices associated with rice terracing can still be observed, especially in Gueday. Important rituals like *begnas* (to synchronise agricultural activities) and *obaya* (rest day) are still performed, and stories and beliefs still constitute the cultural make up of the *i-Agawa*. The stories and beliefs that developed around the practice of rice terracing have served to reinforce the social as well as the economic value of the *payeo* and have also helped constitute and define the identities of the people who share them.

The social practices of building and repairing stone walls

The people of Agawa and Gueday grow rice and vegetables by carving *papayeo* out of the sides of mountains. The sloping nature of upland landscapes requires in some cases the use of soil and water conservation technologies, including stone walling (Brett 1985). Not all *papayeo* are stone walled: Florendo and Cardenas (2001) classified *papayeo* according to shapes, sizes and the type of soil in the area; these determine whether they have to be stone walled or not. In some areas, the type of soil can withstand water and earth pressure (Florendo and Cardenas 2001). In other cases, a *kabiti*, the local term for stone wall, is built to hold the *payeo*, contain the water and, in general, prevent erosion. Stone walls may also be used to increase the area of the *payeo*. For this reason, the goal is to build a properly constructed *kabiti*, one that is sturdy and stable, lasts long and is not prone to erosion. A properly constructed stone wall allows the farmer to deal with the day-to-day demands of rice growing without having to worry about its stability, especially during the rainy season when the walls become vulnerable to erosion. Repairing a *kabiti* is becoming more expensive, especially for owners who are not competent stone wallers. In certain cases, the owners need to hire a group of *kabiteros*, or stone wallers, to do the reconstruction.

The cultural importance of stone walling

All the *papayeo* in Agawa and Gueday need to be stone walled. The stone wall protects the *payeo* that provides sustenance to the family. There is, thus, a compelling need for the *payeo* owner to build a *kabiti* that would last for a long time. To be able to do this, the stone waller keeps closely to the basic rules of stone walling (for example, positioning of stones, the right shape and backfilling) as well as being careful about the *inayan* or taboos that are associated with the practice. Prestige comes with the construction of a sturdy wall, and a good stone waller is acknowledged by the community.

Stone walling is a major element in the culture of the Cordillera peoples. It is employed not only in holding *papayeo* along mountain sides, but also in supporting houses, roads and irrigation canals. It has also been widely adapted in urban centres. Modern-day stone walling involves the use of cement to keep the stones together, but the stone wallers in the two villages do not use cement, they rely on the correct positioning of stones, proper back filling, use of foundation stones, and knowledge about the location to ensure that the stone wall will hold and last.

If one is to use the presence (or absence) of stories, myths or legends as an indication of the importance of a practice, then there is no question of the importance of rice terracing and stone walling to the people. The anthology gathered by the faculty of Besao Elementary School (2002) is replete with stories about the payeo and the kabiti (see also Cordillera Schools Group 2001). One popular story is about a stone wall that kept eroding. The farmer who owned the pond investigated and found that his pond had become a playground of nymphs, known in the village as talaw. He crawled toward the pond and picked up one pair of wings. When the nymphs decided to leave, one of them realised her wings were missing. The man suddenly appeared, startling the nymphs who flew away leaving the other nymph behind. She later married the farmer who refused to give her back her wings. They had two sons. One day, the nymph found her wings and decided to return home. She tearfully bade goodbye to her two sons then left, never to return. There are also stories about rice paddies that were built by the rainbow Balikungan; the i-Gueday point to two rice paddies they claim to have been built by Balikungan.

Since stone walls add to the allure of the *papayeo* for visitors, especially if no erosion is visible, this leads one to ask whether the stone wallers follow some aesthetic standards in their construction. All the stone wallers interviewed said that in building walls, beauty (*pintas*) or appearance (*buya*) is not important. Foremost in their minds during construction is the proper placement of stones to ensure the wall's longevity.

In general, the practice of stone walling contains a range of knowledge from construction to maintenance and repair. These categories of knowledge include *classifying/defining* (types of stones, whether small or big, round or irregularly shaped, hard or soft; kind of soil, whether hard or loose; quality of land, whether stable or sinking), *explaining* (why stones break; why stone walls erode), *estimating* (height of stone wall and area of rice paddy, number of stones to be used), *decision-making* (positioning of stones; what stones to use for a particular space; inclination of stone wall; kind of backfill – pure soil/combination of soil and pebbles), and *trouble-shooting* (solving specific situations or problems; on the spot adjustments/decision-making – e.g., presence of water at the base of the wall). It also includes knowledge about other things, important in the culture (beliefs and customary laws about stone walling; stories and rituals).

Stone walling is a gendered practice as it is the men in the community who are said to possess this knowledge; they are the traditional stone wallers. The respondents said that every male member of the community should learn how to build stone walls; every man should be able to build or repair his stone wall when necessary.

However, community members said that not every male farmer has this skill, that the skill and knowledge of stone walling are something that a person is born with. Using the term *'nailinya'*, they suggest that one is destined to be a good stone waller and that "'you either have (the skill) or you don't'. Although anybody can try to be a stone waller, and some aspects of stone walling can be learned, there are those who are really good at it. Respondents point to a group of men who they think are the highly competent stone wallers in the village, respected in the community. Even among these highly regarded stone wallers, some are seen to be more knowledgeable than others, making it a differentiated knowledge as well. For instance, some of these stone wallers are able to combine aesthetics with function, building durable stone walls that are also pleasing to look at.

There is no age requirement in learning the skill. One respondent learned it when he was twelve years old, another when he was eighteen, and still another when he was already 40 years of age. All of them learned it by trial-and-error under the guidance of an *amam-a* (elder).

Materials

The materials used are of varying kinds. The most important are the stones, and the most important of these according to the stone wallers is the *pegnad*, the foundation that provides footing for the stone wall. The *kinnitoy* are pebbles, placed to fill gaps, which help position bigger stones and allow drainage to counter pressure exerted by water. The *datil* is the last layer of stones, enclosing the *payeo* and serves as a footpath. Protruding from the wall are stepping stones called *pasawsaw*, which allow cleaning and weeding, and movement between *papayeo.* All these make up the *kabiti*, the stone wall that holds the water, prevents erosion and increases the area of the *payeo*. The *abab* is the backfill, compacted soil that holds the planted area of the *payeo*. *Panad* is the base of the *payeo*, holding the soft mud *lumeng*, the topmost layer that is planted with rice. An important part is the *gusingan*, the water outlet which regulates the water in the *payeo*. A pine stick called *tata* serves as boundary and demarcates ownership, and the *dango*, a piece of pine or rattan wood, indicates that the appropriate ritual has been done on the *payeo*.

The processes

In building the wall, the respondents said they followed a series of steps.

Step 1: Position the foundation.

The general rule is to dig deep until the hard part of the earth is reached. Only then are the stones for the foundation positioned, "three levels of stones below the soil surface" (Florendo and Cardenas 2001). Stone wallers agree that these have to be substantially bigger than the stones used for the wall above, so that the foundation can properly hold the wall. However, one of the respected stone wallers in the village believes a huge stone cannot be used if its use does not feel right, 'no haan na nga rebbeng'; it is more important how the stone sits in relation to the soil and how it 'bites' with the other foundation stones.

Step 2: Place pebbles/small stones.

The same stone waller believes that secret of a sturdy stone wall lies in putting pebbles at the base of the foundation stones to serve as drainage for water. This helps counter the pressure exerted by water on the wall.

Step 3: Position individual stones.

Starting from the foundation stones, individual stones forming the wall are carefully positioned until the desired height is reached. The foundation stones are not visible as they are buried below the wall. During the positioning of individual stones, specially selected stones protrude strategically so that farmers can step on them when working on the wall.

Step 4: Compact the backfill.

Concurrent with the placement of stones is the compacting of the backfill. The backfill serves to extend the area of the paddy. It needs to be compacted properly to avoid water from building up, otherwise the pressure of water in the backfill would eventually 'push' the stone wall.

Step 5: Placing the water outlet.

In certain cases, it is necessary to decide early where to put the water outlet. This allows extra care in the selection and positioning of stones directly below the outlet, which is one of the weaker portions of the wall as it is subjected to the constant flow of water.

Step 6: Finishing the wall.

The last step is the placement of *datil*, the last layer of stones in the wall.

Some technical knowledge involved in building stone walls

Among the skilled knowledge in building the stone walls are the following.

a) Positioning individual stones

Respondents agreed that the careful positioning of individual stones determines the longevity of the stone wall. The most important lesson in positioning a stone is maximising its contact with the other elements (i.e. with adjacent stones and with the backfill). To help position individual stones, pebbles and smaller stones are placed between gaps. The rule is to 'position the bigger stones properly making sure that they fit snugly with respect to the other stones and the backfill, and support these with pebbles or smaller stones'. They say being able to discern the right position is an important skill that every stone waller should have.

The following are selected insights by some of the acknowledged stone wallers (referred to here as S1, S2 and S3) regarding the positioning of stones. These form part of the content of knowledge that they learned from older stone wallers and from their experience, and are what they teach to those who are learning the skill.

> S1: Make sure that each stone fits snugly with the other stones, so that even small ones may be used, not only big stones.

> S1: It is important for the stones to fit properly, for if three stones start moving, the rest will eventually follow.

> S1: It is alright even if the appearance of the stone wall is not so good as long as the stones are properly positioned and the backfill is compacted well to ensure that the soil will not be washed out.

> S2: The stones need to 'bite into each other' or 'tighten up like screws'.

S3: Do not position stones like the kernel of a corn. This will cause the wall to erode easily.

S3: Even if most of the stones fit properly, if there is just one stone that is misplaced, this stone will eventually pull those stones above it.

b) Rotating the stones

A group of stone wallers doing repairs on an eroded *kabiti* led by another elder (S4) spoke of the different ways of rotating a stone in order to find its proper position:

- *Dampagen*: 'setting up a flat stone on its side', positioning thin flat stones with the longer cross-sectional dimension placed vertically to maximise contact with the other stones. Respondents S1 and S3 said another reason why this positioning is preferred is that there is less space for weeds to grow.
- *I-paungdos*: positioning stones with a longer dimension as headers instead of stretchers, which according to S1 prevents unbonded alignments (Conklin 1980).
- *I-suni*: rotating other stones continually until the 'best fit' or 'right feeling' is achieved. This is one of the common expressions used by most of the wall-builders when asked to describe how to position a stone: *suni-em*, which literally means, '(you) rotate'.

c) The 'rightful stone'

The concept of a rightful stone implies a set of criteria that a stone must possess for it to be considered appropriate to fill a space. In general, the criterion that needs satisfying depends on whether the shape allows for more contacts between elements and ensures greater stability under conditions of weathering, seepage and ground movements (Conklin 1980). For this reason, the stone wallers claim that angular and irregularly shaped stones are preferred, for they provide greater contact with the other elements.

d) Assessing the stones

Once chosen, a stone is seldom replaced. The following possibilities may happen:

- It fits the space so that it has maximum contact with adjacent stones and the backfill (*abab*). This is not accidental. Normally, the stone waller has an 'eye' or a 'feel' for the space that needs filling up. As he scans the pile of stones, his choice of the 'rightful' stone depends on his perception of the space in question; the selection is a result of conscious matching of 'chunks'.

- It is rotated several times until the placement gives the right 'feel'; only then does it confirm its being a rightful stone. This involves a sequence of rotations and trial and error before the 'best fit' is achieved. Rotating a stone several times before it is placed makes sense cognitively; Kirsh and Maglio (1994) show in their study of the game Tetris that players over-rotate falling shapes to see the shapes in more than one orientation, which leads to faster and more accurate placement decisions. They see the action of over-rotating the shape as part of what they call *epistemic actions* that people take to simplify their internal problem-solving processes.
- In certain instances, the stone waller may opt to reshape the stone by chipping it, using another stone, to satisfy the above criterion. This involves estimating the shape of the stone that is needed for the space.

Height and shape of a wall

The highest *payeo* observed in Gueday was close to 4.3 metres, although residents claimed that there was a 6-metre high *payeo* near Lamagan, north of Gueday. Generally, the shape of a wall varies from vertical to a slightly concave batter as this holds best. A wall that bulges with the hill's pressure is not desirable.

The respondents are aware of the vulnerability of high stone walls, and certain techniques have been developed to mitigate the possibility of erosion. Two of these techniques are described below.

a) Putting in a secondary support (segunda):

The wall is divided into two parts as shown in Figure 3.1 The lower part, called *sumkad*, provides support to *dumteg*, the main wall, which is made vertical (*payeo* A) or inclined relative to the vertical (*payeo* B). Note the *Payeo* B has a bigger area than *Payeo* A. Not shown in the figure is the back of the wall, which generally follows the contour of the mountain.

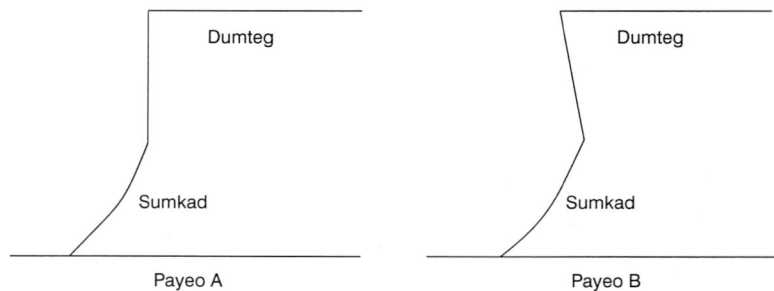

FIGURE 3.1   Sketch of a high wall with a secondary support

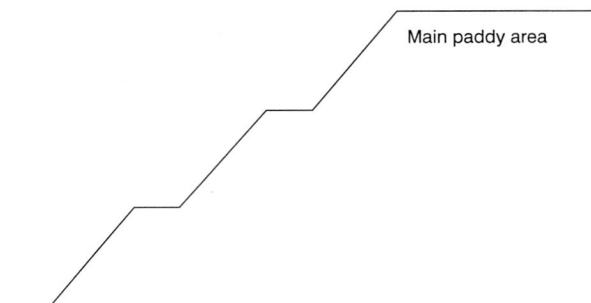

FIGURE 3.2 Sketch of a high wall with secondary terraces

b) Terracing the wall

The second technique for high stone walls is to 'terrace' it (*sanipat*). The respondents claim that this is stronger and makes for easier cleaning. The small spaces created in the lower layers of the terrace may be used to plant rice as well as sweet potato (see Figure 3.2).

Some mathematical concepts and stone walling

The language of stone walling and mathematics

One possible way to look for mathematical concepts is to study language, how they talk about stone walling, including the use of metaphors that are associated with the practice.

There is no distinction between pressure and force. The local term for both is '*pigsa*', used to describe the pressure/force of water or the soil. Stone wallers refer to *pigsa ti danum* (pressure of water) or *pigsa ti daga* (force of the soil) as concerns in the construction of a sturdy stone wall.

There is no local term for friction. The idea of friction seems to be captured in the way they describe the positioning of stones, that is, when they talk about the stones as 'biting into each other' or 'tightening like screws.' The idea is to maximise contact points between the stones to produce greater wall strength.

Some beliefs reinforce the need to build the wall in a concave batter.[1] For example, they have the belief that 'if the wife is pregnant, the man should not build a stone wall' or that the stone waller should not have sex while building stone walls for this might cause erosion of the wall. Stone wallers predict that a wall that resembles the (convex) shape of a pregnant woman or one with a full stomach will erode in due time.

Just as metaphors are important in mathematics (Lakoff and Nunez 2000), stone wallers use metaphors in talking about a 'rightful' stone that should occupy a particular space in the wall – the stone that provides the 'best fit', a stone that 'feels right'. A durable stone wall is used to describe the rearing of a child or the quality of a relationship between two people. A child with good values is like a stone wall with a strong foundation, while a lasting relationship is like a sturdy stone wall, it has permanence. A sturdy stone wall stands the test of time, much like an elegant mathematical proof.

Parallelisms with mathematics

There are several parallels between mathematics and stone walling.

Practices

Stone walling practices show parallelism with some practices in mathematics. While the QRS ideas (see above) are reflected in the sizes and shapes of stones, in the considerations of height and slope of the stone wall, in the area of the *payeo* among others, certain processes and practices resemble those in mathematics. For example, the practice of stone walling is a systematic body of *knowledge* that incorporates the social, cultural and the analytical. It includes the knowledge of the materials used, the stones and the soil. And there is a myriad of *competences* associated with stone walling, foremost of which is the selection of the right kind of stone for the different uses – foundations, top layer, pebbles and stepping stones. But in particular, there are how the stone wallers are regarded by the community, how they communicate and argue about the practice, and how the practice itself adapts to changes.

The practitioners

One parallelism between stone wallers and mathematicians is in the way they are regarded by their respective communities. There is the prevailing idea among non-mathematicians that one is born with the skill and knowledge to become a good mathematician; the same idea exists about the good stone waller in the community. The parallelism extends to the status that is accorded a highly competent mathematician and a good stone waller: the community generally considers a mathematician intelligent, while a good stone waller is generally a respected member of the village, and most of the time is considered a wise person. They are expected to mentor the young about the practice.

Talking about the practice

It is also important to note how stone wallers talk about stone walls and walling *ex situ*. Stone wall construction problems are discussed not only on the site but also in their homes or the *dap-ay*. In one interview, the respondent used pebbles to demonstrate the proper positioning of stones. Another stone waller, when interviewed in his house, used bottles of soft drinks to show the importance of placing stones as 'headers'. These encounters suggest that stone wallers sometimes use representations and models in talking about the practice. Several group interviews showed that the bases for their judgments and arguments about matters pertaining to stone walling are generally agreed upon and accepted by them. While it is clear that the stone wallers were not creating mathematical knowledge in the process, the ways that they think, talk and go about the practice parallel the ways mathematicians think, talk and go about solving some mathematical problems.

One difference is the way that the knowledge is recorded and communicated. Agawa and Gueday are societies that are steeped in oral tradition; knowledge is passed on from generation to generation orally and through ritual. Mathematical knowledge is also rich in tradition, but much of the success of this knowledge is due to the written text that makes use of the symbols and grammar of the mathematical language.

A commonality is the use of trial and error as learning and teaching strategies. In stone walling, this teaching and learning technique is central. In mathematics, trial and error methods have a place in mathematical instruction.

The practitioners also make use of metaphors, stories and myths in communicating the knowledge about stone walling, reinforcing knowledge in the community (Alangui 1997; St Clair 2000). The metaphor of the pregnant woman or the full stomach already mentioned communicates knowledge about the desirability of a concave batter in the construction of a new stone wall, while an old *payeo* that starts to take the shape of a pregnant woman is soon to erode; such metaphors reinforce knowledge of what a sturdy stone wall should look like. Existing techniques in building a sturdy stone wall (like proper positioning of stones) are informed by this knowledge. This is akin to the way mathematicians might refer back to the mathematical literature in order to illuminate a theory or a concept or to develop new mathematical knowledge (cf. Alangui 2010). In both instances, practitioners invoke forms of 'records' and 'recordings' of their respective knowledge to inform current practice, in much the same way that numeracy events may be photographed (Baker 2008).

Conclusion

Using Reckwitz's (2002) concept of social practice, the building of stone walls may be considered a social practice. As a highly systematised practice that has persisted for hundreds of years in the villages of Agawa and Gueday, the practitioners have developed competences in dealing with the materials (stones, soil, other materials) needed in the construction. And because of the context and culture associated with stone walls, these materials and competences assume symbolic meanings and embody shared aspirations between the stone wallers and the larger community.

Several of the steps in stone walling may be considered as numeracy events: the choice of stones, deciding which stone is appropriate for the different parts of the wall; placement of stones; decisions on height and shape of the wall;

and actual construction. All these events may be seen as comprising numeracy practices: the construction takes place within a cultural context and ideology, with meanings, symbols and values that are important to the community. The process of positioning an individual stone, whether it is part of *pegnad* (foundation) or *kabiti* (the main wall), involves a series of decision-making that takes into consideration important constraints.

The stone wallers do not count the stones needed in the construction. No computations are involved with respect to the height and shape of the wall. At best, they estimate, they model, make predictions. They are also creative, for example, breaking down a big stone to produce smaller stones that could serve various roles in the wall. The numeracy embedded in the practice of stone walling is a type of numeracy without numbers. It is also a useable numeracy in the sense of Kanes (2002), one that is clearly used in real-life problem solving.

What this chapter shows is that it is possible to go beyond the traditional confines of numeracy, beyond the classroom and beyond numbers, to include the investigation of highly systematised cultural practices and the numeracy embedded in such practices. I see two implications of this chapter. One is related to the theorising of numeracy as knowledge embedded in cultural practice, situated in context and as social practice. The other implication is in mathematics and numeracy education. How do we take advantage for example of the mathematical/numeracy knowledge embedded in stone walling to inform the mathematics curriculum? How can mathematics and numeracy educators use the knowledge about estimations, modelling and prediction, about shapes, height and area, and spatial relationships that are embedded in stone walling construction to enrich and contextualise lessons for students and adult learners? Given the diversity and wealth of cultural practices, the possibilities for developing culturally relevant mathematics/numeracy lessons for the communities who own such practices are endless. Conceptualising numeracy as a social practice allows these possibilities.

Note

1   Batter is a technical term in architecture and building to refer to a wall of a particular shape or slope.

References

Adam, A., Alangui, W. V. and Barton, B. (2010) "Bright lights and questions: Using mutual interrogation" *For the Learning of Mathematics* 30(3) 10–16.

Alangui, W. V. (1997) *Indigenous Learning Systems in a Kankana-ey Community (Mt. Province)* Quezon City, Philippines: Education Research Program, Center for Integrative and Development Studies, University of the Philippines and the Department of Education, Culture and Sports, Bureau of Non-Formal Education.

Alangui, W. V. (2010) "Stone walls and water flows: Interrogating mathematics and cultural practice". Unpublished thesis, Department of Mathematics, The University of Auckland.

Baker, D. A. (2008) "Using sand to count their number: developing teachers' cultural and social sensitivities." Retrieved 4 April 2017 from http://www.waalc.org.au/09conf/docs/Using_sand_to_count_Baker.pdf.

Barton, W. D. (1996) "Ethnomathematics: Exploring cultural diversity in mathematics". Unpublished thesis, Department of Mathematics, The University of Auckland.

Bauzon, L. (1999) "A conceptual framework for the study and teaching of Philippine history and for nation-building" paper presented at the Second National Conference on the Teaching of Philippine and Asian History and Culture, 12–16 April, University of the Philippines, Diliman.

Bodner, C. C. (1986) *On the Evolution of Agriculture in Central Bontoc* Columbia, MS: University of Missouri-Columbia.

Brett, J. P. (1985) "Stone walls and waterfalls: Irrigation and ritual regulation in the Central Cordillera, Northern Philippines". In Hutterer, K., Rambo, T. and Lovelace, G. (eds), *Cultural Values and Human Ecology in Southeast Asia* Center for South and Southeast Asian Studies, Ann Arbor, MI: The University of Michigan Press.

Conklin, H. (1980) *Ethnographic Atlas of Ifugao* New Haven, CT: Yale University Press.

Cordillera Peoples' Alliance (2003) *Primer on Cordillera Peoples and Indigenous Peoples' Rights* Baguio: Cordillera Peoples' Alliance.

Cordillera Schools Group (2001) "Stories". Unpublished manuscript. Baguio: Cordillera Schools Group.

Fiar-od, C. B. (2001) *Besao Traditional Knowledge on Spiritual Beliefs: Its Contributions to Sustainable Development* Bontoc, Mt. Province: Mountain Province State Polytechnic College.

Florendo, M. N. and Cardenas, M. (2001) *Towards the Indigenization of Formal and Non-Formal Education in the Cordillera: Cordillera Rice Terracing as an Agricultural System* Baguio: University of the Philippines.

Institute of Environmental Science for Social Change (2003) *Besao Ancestral Domain Management Plan* Quezon City: Institute of Environmental Science for Social Change.

Kanes, C. (2002) "Towards numeracy as a cultural historical activity system". In Valero, P. and Skovsmose, O. (eds), *Mathematics Education and Society, Part 2. Proceedings of the Third International Mathematics Education and Society Conference, MES3, 2–7 April 2002,* Helsingør, Denmark. Roskilde: Centre for Research in Learning Mathematics, The Danish University of Education, Roskilde University, Aalborg University.

Keesing, F. (1962) *The Ethnohistory of Northern Luzon* Stanford, CA: Stanford University Press.

Kirsh, D. and Maglio, P. (1994) "On distinguishing epistemic from pragmatic action" *Cognitive Science*, 18 513–549.

Lakoff, G. and Nunez, R. (2000) *Where Mathematics Comes From* New York: Basic Books.

OPSRA (1995) *Operationalization Plan of the Social Reform Agenda: Minimum Basic Needs, Municipality of Besao* Besao: Municipality of Besao, Mt. Province.

Preston, D. (2003) "Changed household livelihood strategies in the Cordillera of Luzon" *Journal of Economic and Social Geography*, 89(4) 371–383.

Reckwitz, A. (2002) "Toward a theory of social practices: A development in culturalist theorizing" *European Journal of Social Theory*, 5(2) 243–263.

Scott, W. H. (1966) *On the Cordillera: A Look at the Peoples and Cultures of the Mountain Province* Manila: MCS Enterprises, Inc.

Scott, W. H. (1974) *The Discovery of the Igorots: Spanish Contacts with the Pagans of Northern Luzon* Quezon City: New Day Publishers.

Sefa Dei, G., Hall, B. and Goldin Rosenberg, D. (eds) (2000) *Indigenous Knowledges in Global Contexts* Toronto: University of Toronto Press.

Spencer, J. E. (1952) *Land and People in the Philippines: Geographic Problems in Rural Economy* Berkeley, CA: University of California Press.

St Clair, R. (2000) "Visual metaphor, cultural knowledge, and the new rhetoric". In Reyhner, J., Martin, J., Lockard, L. and Gilbert, W. S. (eds), *Learn in Beauty: Indigenous Education for a New Century* Flagstaff, AZ: Northern Arizona University.

Staff of Besao Central School (2002) *A Compilation of Besao Folktales, Legends, Actual Events and Stories about Local Heroes as Supplementary Reading Materials in English for the Infusion of Values* Besao, Mt. Province: Department of Education, Cordillera Administrative Region, Division of Mountain Province.

Tauli-Corpuz, V. (2001) "Interface between traditional religion and ecology among the Igorots". In Grim, J. (ed.), *Indigenous Traditions and Ecology: The Interbeing of Cosmology and Community* Cambridge, MA: Center for the Study of World Religions, Harvard Divinity School.

4

'TEAR IT OUT AND RIP IT UP OR YOU MIGHT GET CHARGED AGAIN'

Paying debts at the company store in a farm workers' camp in Mexico

Judy Kalman and Diana Solares

Introduction

Some of Mexico's most important agricultural industries, including grains, fruits, and other crops, are widely cultivated in the northeastern states of Sonora, Sinaloa, Baja California, and Baja California Sur. Migrant laborers from the southern states of Mexico are hired every year to cut and pack the wide variety of products from that region, ranging from grapes, tomatoes, asparagus, wheat, and corn to beans, oranges, watermelon, potatoes, alfalfa, and cantaloupe (Moreo Gastelum 2015).

Migrant workers come from some of the country's poorest and most marginalized communities, especially from the states of Guerrero, Oaxaca, and Veracruz. Once contractors hire them, they are taken by bus, often with their families, to large agricultural fields, located as far as 45 hours away from their hometowns, with a promise of a salary, room and board, and transportation (Muñoz 2012). However, on arrival, what these families find is very different from what was promised to them. Before work has even begun, they are often saddled with debt to contractors for their travel and food costs during the trip. Workers' camps, located far from urban centers, offer housing, a company store, and transportation in exchange for fees established by employers. After long days in the field, an entire family's efforts—children, teenagers, and adults—add up to less than minimum salary (Maldonado 2005).

In this context, calculating and keeping records of produce collected in the fields and debts from purchases at the company store can become a potential source of negotiation and conflict for these families. In a recent study, Solares (2012) analyzed the production of these and other numerical records from several agricultural areas in Mexico's north (checks, receipts, worker's daily production, and other documents). Numerical information is produced at specific moments and circulates from party to party, each one interpreting and using it according to their interests.

This chapter discusses the conditions in which notes, calculations, and bills are produced (Blommaert 2005, 2008)[1] and the ways these conditions shape how these activities take place and how, at the same time, they are part of a greater context of poverty and inequality. In other words, in situations of extreme inequality, such as the ones reported here, inequality materializes in many ways, and documents displaying numeric information generated through diverse activities are also evidence of these interactions and conflict. Under what conditions are record-keeping strategies and calculations for monitoring debts produced? How do these conditions affect strategies? How do participants justify and explain their strategies? How do participants solve tensions that arise as part of these transactions?

Those who run the company stores are employed by the owners and are the ones charged with keeping track of working families' debts, and they are known for overcharging them. The storekeepers' excessive prices and credit systems contribute to making working families captive debtors. As documented, a large portion of family income goes to paying debts at the company store; uncontrolled family expenses can exceed income, and if they do not settle their bills, they are not allowed to leave (Barrón 2012).

For this reason, Ana and her family, like many other families, keep track of their debts and the crops they collect in the fields in Caborca, located in northern Mexico. Ana, a bilingual Nahuatl-Spanish speaker, comes from the state of Guerrero, about 2500 kilometres away. While her adult children worked in the fields picking grapes, she looked after other workers' young children, taking care of them while their parents were in the fields. We chose this family as a telling example because the ways they interacted and collaborated around keeping track of their expenses and debts became especially visible in the data. During picking season, work in the fields begins before 6:00 am, other members of Ana´s family along with other laborers are bussed from the camp where they are housed to the fields. Their day ends at 3:00 pm when they return by bus.

Ana did not go to school and never learned to read or write,[2] and while she relies on memory to keep track of her accounts, her eldest sons and daughters are responsible for reviewing the storekeeper's records in a small notebook he provides for them. At the end of the week, her adult children check the notebook ('el librito', the little book, as she calls it) to add up the bills and pay them with the same check they receive from their employer. Ana constantly worries that her children are paid fairly for their labor and that the store does not overcharge them. In her hometown, back in Guerrero, Ana operated a market stall and was renowned among her peers for her mental calculation abilities.

Working in the fields, far away from home, Ana finds herself facing great uncertainty, distrustful of those in charge, and at a loss when facing the storekeeper's calculations and relies on those who have been to school to mediate for her: to interpret and revise written accounts, to identify discrepancies, and sometimes to confront the storekeeper, even though she skillfully engaged in tallying bills and engaged a variety of resources (such as making mental calculations, memorizing amounts, or paying for items one by one) to solve similar situations that required accounting skills when she worked in the market in her hometown. The way Ana sees herself regarding calculation skills changes radically in these two contexts.

When we examined the different ways of recording, calculating, and paying bills in the fields near Caborca, Sonora, and analyzed the contexts for annotations and calculations, we found that indeed, interactions and conflicts between participants have strong effects on the ways this information is produced. Other factors can influence these conditions (for example, how information is recorded, where it is recorded, how and where it is calculated, who is responsible for recording and calculating). We have verified that inequality is a relevant element in production conditions; inequality permeates numerical records and calculations when conflict or disagreements emerge between participants (clients and storekeepers, when it comes to paying debts; workers and employers, when it comes to paying salaries). Therefore, interactions and conflicts that arise regarding certain numerical documents are often present in the documents themselves.

Closely studying and documenting literacy practices began at the end of the 1980s with NLS (new literacy studies), and quickly spread as a theoretical and methodological approach that focused on the specificity of reading and writing in diverse contexts (Canieso-Doronila 1996; Collins and Blot 2003; Cope and Kalantzis 2000; Heath 1983; Tett, Hamilton and Hillier 2006; Wagner 1993). It has enabled the description and analysis of situated practices, paying special attention to the heterogeneity of ways of doing and signifying, to bilingualism, and to vernacular writing. The tradition has extended to Latin America studies: Wogan (2004), Kalman (1999), Zavala (2002) and Castanheira (2013) for example have revealed a variety of practices, representations, and uses for reading and writing. The same perspective has also been used to focus on specific numeracy practices.

Mathematical practices (including numerical practices) have been studied under a variety of theoretical and methodological perspectives, for example, Lave's (1982) study of adults performing calculations in a supermarket or Carraher, Carraher, and Schliemann (1995), who analyze and compare the ways children in Brazil make calculations when selling in the streets with the way they make calculations at school, in classroom situations. Recent studies have been conducted in Latin American countries by authors such as Delprato and Fuenlabrada (2012), De Agüero (2006), Soto (2001), Knijnik (2003), Padilla (2015), Broitman (2012) to name a few. This research has looked at maths in daily activities under specific circumstances and determined purposes; for example, artisans selling their product and determining how to distribute earnings, painters who establish the cost of their work, field workers who calculate their proportion of crops, homeless families measuring surfaces in a particular way, child workers who estimate storage capacity for water selling, illiterate adults who employ arithmetical knowledge at work. These studies have used different theoretical and methodological perspectives, but they coincide when recognizing maths as a social and cultural practice and when approaching mathematical knowledge used in activities within their subject's natural setting (Solares 2012).

In the following pages, we will analyze the way Ana and her family record and calculate their debt with the local store. This chapter has three sections; the first section will review a few theoretical concepts that have shaped this research, in particular, notions of literacy, numeracy, social practice, event, and activity. This section will also provide a brief but necessary overview of our fieldwork and the focal informant selected for this chapter. The second section will present the strategies employed by the subject to confront the storekeeper's charges and the verification criteria that uphold them, and we will analyze specific examples of how these criteria are used; the third section provides a few final comments.

Mathematics in reading and writing practice: new literacy and numeracy studies

The publication of *Literacy in theory and practice* (Street 1984) sparked a debate around the instrumental nature of reading and writing, and by extension of mathematics (Kalman 2008; Kalman and Street 2009; Moss 1994; Reder 1994). While written forms of representation, signification, and interpretation include an operative handling of certain codes, in his seminal book, Street states that the use and dissemination of written language and mathematics are not neutral or autonomous. He claims that all the ways of doing, using, signifying, and interpreting through writing are deeply ideological, situated in specific contexts and linked to broader political and cultural agendas. While readers and writers require both technology and skill to participate in literacy events, they also engage relevant social knowledge, mediating tools for producing and comprehending texts. This is what allows them to read and write within a specific context, within social and power relations (Scribner and Cole 1980; Heath 1983).

In this sense, literacy[3] is more than learning the basic skills of reading and writing (Kalman 2001, 2004, 2016). Literacy is a situated practice that implies the ability to read and write to some degree, but also engaging with other people through and about texts. In this sense, literacy is as much a social phenomenon as an individual one. Barton and Hamilton (1998) define literacy as a series of social practices that can be deduced from events. These practices are historically constructed and are shaped by social institutions and power relationships, which is why some literacy practices can become more visible than others. Dyson (1997) proposes that being literate means being able to use reading and writing to participate in socially recognized cultural events and to establish and maintain social relationships.

Several researchers have defined literacy practices as the unit of analysis in studies such as these; that is, the cultural ways in which people engage with written language in their daily life (Barton and Hamilton 1998; Kalman 2001; Street 1995). Literacy practices involve values, attitudes, emotions, and social relationships. As part of their research, these authors distinguish between literacy events and literacy practices: the notion of *literacy event* is used to identify and analyze activities where written language plays a central role. 'Events are observable episodes which arise from practice and are shaped by them', and this notion stresses the situated nature of the activity (Barton and Hamilton 1998, 7). Events are shaped by written text, but physical texts do not have to be physically present; they can simply be recalled and still play a central role in social interaction.

While some authors tend to use 'event' and 'activity' as synonyms (as Barton and Hamilton do above),[4] in this chapter we will use the term 'activity' to describe and analyze situations where written language—specifically written numbers—plays a central role in the fields, mainly because the notion of 'activity' implies reaching a *goal* under determined conditions, while 'event' is more of a sociolinguistic term concerned with uses and functions of language. Social practice is a broader notion of patterned activities, an abstract term, that implies both what is observable (actions and speech) as well as what can be inferred or reported (attitudes, beliefs, and ideologies). Practices are made up of multiple activities. Within the concept of activity, the goal and the ongoing conditions shape actions and permeate its conclusion. In this case study, our main interest lies in the relationships between field workers and storekeepers, specifically regarding debt recording, interpretation, and calculation. The tensions that arise around conflicting interests and perspectives are highly relevant to the way certain records and numerical calculation strategies are produced.

We look at the data from a social practice perspective, which becomes visible in situated activities. Our vision assumes a 'reciprocal relationship between people and their social environment' (Valsiner and Rosa 2007, 4). We have structured a framework for interpretation, which enables us to understand the way working families, field supervisors and local storekeepers use written documents, numerical annotations and mathematical thinking as mediators during various negotiations. This theoretical perspective sees all human activities as both *situated* and *contextualized* (Boyarin 1992; Kalman 2002; Levine 1986; Reder and Dávila 2005), while subjects, objects, lives, and worlds are constructed by and through relationships. That is, the contexts of people's lives aren't merely containers or backdrops, nor are they simply whatever seems to be salient to the immediate experience. Persons are 'always ... located uniquely in space and in their relations with other persons, things, practices and institutional arrangements' (Lave 2011, 152). Written and numerical practices are more than just distributing text, operating a calculator, pencil and paper, understanding company rules, and conventional counting and annotating.

The next section describes a series of situations that were observed throughout two months of fieldwork. To collect and analyze this chapter's data and as a way of understanding how numerical annotations are recorded, used and interpreted, Solares (2012) made observations in the fields, in packing areas and camps, with workers from grape and asparagus fields in Caborca, Sonora. She interviewed twelve boys and girls (some of whom were agricultural laborers), seventeen adult laborers (croppers, record-keepers, supervisors and child-minders) and their families (seven families total). Through interviews, she researched different ways of recording each worker's daily production, as well as the way families kept track of and controlled their spending in the local store, as well as how they resolved discrepancies with storekeepers (she also interviewed the storekeepers of both local stores). Families' different verification strategies were also identified as well as the criteria that support each strategy.

Written numbers as part of other texts

Documents written in the fields that display numerical information are produced during certain moments, circulate between different users, and are interpreted according to users' work-related functions, position, and interests. Some of these documents are also linked to each other through the diverse activities and interactions that take place around their content.

These interactions and activities can be discerned from an explanation given by two youngsters regarding worker's checks paid in a field in Caborca, Sonora, and how this is connected to family debt records. Silvino says that checks are cashed at the local store, but only after workers have paid off their debts at that same store. Thus, they must hand in their debt notebook as well when cashing their check. Young Adela, on the other hand, states that the way families control the debts that they already have paid is by tearing out the sheets where they had tallied up their debts to avoid being charged more than once:

Silvino: You change it [the check] at the store … if you have any debts […] they will give you money [referring to money leftover once debts are paid] and you will change the check … they will give you money, but you must take the notebook […]
Interviewer: On Saturdays, payday, what happens to the notebook?
Adela: You take it in to get paid and they erase it …
Silvino: They tear it up.
Adela: The little piece of paper in the notebook, one of those little notebooks, um, you tear it out and tear it up, because if you don't you will make a mistake … they will charge you again.
Interviewer: So you tear a page out each Saturday?
Adela: Yes.

Debts are recorded in two notebooks, one kept by clients (families) and the other kept by storekeepers. Both documents display numbers but do not name products. Those figures represent money, but do not distinguish whether they represent the price per unit of a given product or if they are the total for various products. Neither accounting includes a date, and the only number that doesn't reflect an amount of money is the one used to identify clients (#185, for example). Both storekeepers and families must keep track of debts and payments, and the main way to do that is for each party to use their notebooks. When debts are paid, storekeepers cross out the record in their notebooks while families tear the corresponding page out of their respective notebook and destroy it, as Adela explained. It is important to point out that families do not write down debt records in their notebooks, storekeepers do.

The most evident *numerical activities* have to do with the social practice of accounting, specifically tallying the bill of groceries bought on credit. The principal activities are visible in registering, calculating, and paying debts, the storekeeper's annotations in both notebooks, and the way they punch in numbers when using a calculator to add up debts each time a client comes to pay. What remains invisible are the calculations performed by families in their rooms or camps the night before paying, and the way they control the daily debt record and the amount they owe, especially for those who cannot read or write.

What kind of resources can they rely on to keep track of their debts, when most of the adults in the family see themselves as non-literate? More specifically, how do families know that the numbers in the storekeeper's notebooks coincide with the numbers in their notebooks? And more to the point, how do families confront storekeepers when differences and conflict arise?

As previously noted, Ana, the focal informant for this chapter, comes from the state of Guerrero and said she never went to school. She explained that she could not read nor write and therefore asked her older children to look over their notebook to ensure they were not charged more than they owed. These efforts took place in two stages, the first when revising debts acquired in a single day, and the second when they did their accounting on the weekend. Ana's discourse reveals some of the strategies she employs to keep track of each day's debts.

Like Ana's case, in other families, the children who were more accomplished readers and writers took on leading roles when reviewing records and confronting storekeepers if differences arose. Families tended to gather the night before going in to settle their debts to review their annotations and to add up the amounts registered in their notebooks with the help of mobile phones and calculators. When bills were short, they might rely on mental maths or written calculations, though only one of the interviewed families claimed to use this resource. Storekeepers, on the other hand, kept records of their own, used calculators and some written mathematics for tallying accounts.

When there are differences, storekeeper's and families' notebooks were compared. However, it is the storekeeper who wrote in both documents. Therefore, families had to pay careful attention at any time when debts were being recorded, and not just during weekly accounting in the stores.

How did Ana and other non-literate adults face this situation when their children were not available to oversee the storekeeper's debt calculations?

We identified four strategies[5] that Ana used to keep track of her expenses. However, it is important to point out that these strategies are not used only by unschooled and under-schooled adults, some of those interviewed were conventional readers and writers but still kept track of their expenses and debt in ways similar to Ana's. This suggests that these strategies have more to do with writing down purchases and reconciling accounts in conditions of social inequality that are a common thread throughout many different agricultural fields in northern Mexico and with the degree of poverty that looms over working families.

Strategy 1: Relying on other people and on memory

Because Ana provided child care for young children, the parents could work in the fields. Some of those families left their debt notebooks with Ana in case she needed to buy something for their children. At times, she was in charge of up to three notebooks. To do so, she used her memory to keep track of which products she purchased and where each transaction was recorded; then, when Ana's grown children returned from the fields in the afternoon, she asked them to be sure that the amounts recorded matched the product she had purchased.[6] If these did not match, Ana would go back to the store immediately to make the claim and would rely on her memory (remembering what she bought) and the amounts that her children read to her. For example, she commented that on one occasion she was charged 81 Mexican pesos for juice (approximately 20 times its original value). After her daughter read the amount to her, Ana immediately headed straight to the store to complain.[7]

Ana:	I tell them [her sons and daughters-in-law] to check if they wrote it down right where I bought it, sometimes the owner writes things wrong […] sometimes they write the wrong thing, charges more, sometimes […] then when they [her children] come home [from work] I show them, and say: [Check] if they wrote it right, ok, they say 'no' …, the other day, for a juice that costs four pesos, they wrote down eighty-one!
Interviewer:	What! How? That much?
Ana:	uh-huh, eighty-one they wrote, my daughter says, no, mother, she says, it says eighty-one here, okay, so I ran to ask them [the storekeeper], why? Can't you see? I say they're watching television and don't pay attention; they're looking at the television there and writing here. I say, why did you charge me eighty-one pesos for this? I had to complain about this because it says eighty-one, right? So then he erased it and I told him. You charged me too much for a four-peso juice.
Interviewer:	Did he erase it?
Ana:	Yes. He erased it and wrote four pesos instead.

Along with her use of memory and asking others to read the notebook for her, Ana also used the *trust criteria*. Given the worker's experiences with storekeepers, she asked a trustworthy person to review the records and mediate its interpretation for her. The difference between what was written and the known price for a container of juice inspired Ana to return to the store to rectify the records.

Strategy 2: Number interpretations based on memory

Even though Ana said she could not read, she had a strategy for making sure that the number of items listed corresponded with the number of items purchased. She would try to identify the number of amounts written in the notebook (she referred to this as how many 'letters' were written) and how many amounts were added afterward, as a way of assuring that the number of figures matched the number of items purchased.

> To stop them from writing the same product twice, I count [she signals an amount on the list of debts] first this one, if it's one, two or three or four, however many; before entering the store I've already counted, how many letters [Ana´s way of referring to each entry] are written here, how many they are going to write, if it's just one or two …

Ana relied on an *amount comparison criteria* to interpret the list of debts for each purchase; she tried to reconcile the number of amounts [entries] written in the records with the number of products purchased. However, storekeepers often added up the price of several products, which complicated Ana's strategy ('… sometimes, with milk, with meat … with some chilies, green ones, tomatoes, eggs … sometimes they write a hundred and some …'). For this reason, she usually asked the storekeeper for the unitary price of each product and repeated them from memory to her kids as they reviewed the accounts. She didn't always obtain an answer from the storekeeper (according to Ana, this would happen when the store was full of clients).

> If not, sometimes I ask how much this costs, or how much this other thing costs … sometimes he wants to tell me, other times he doesn't, he adds it up quickly, because sometimes there are many people …

Ana's narration explains the criterion that she uses to verify what is written by the owner of the store, even though she does not know how to read or write. It is a valuable resource learned through surviving in conditions of extreme poverty and inequality that allows her to interact with written numerical information and to some extent confront the storekeeper about his written records. However, it is worth noting the limits of this resource when the tally is the sum of several items' prices. In this case, Ana depends on the shopkeeper to tell her the unit cost of each product and, moreover, depends on her ability to remember then this information afterward when she and her children go over the accounts and verify them. Given the above, Ana's strategy is extremely complex, costly concerning the effort, and the tasks involved, and is further evidence of the vulnerable situation in which these families live.

Strategy 3: Estimations based on the cost of merchandise

Ana knew the price of several products, which allowed her to estimate how much she could buy and how much they should charge her.

During an interview, Solares showed Ana a debt notebook where several amounts had been registered. Then she pointed to one of the quantities and asked her what she might have purchased. Ana's response was based on her experience as a client:

Interviewer: Look, for example [...] here is a written number that says sixty-six. What can you get for sixty-six?
Ana: Yes, sometimes milk, with meat ... some chilies, greens, tomatoes, eggs ... sometimes they write a hundred or so.
Interviewer: But when you're getting several things ...
Ana: Yes. Like now, if you get oil and eggs and tomatoes ... it seems that it doesn't take much to add up a hundred pesos.

She also seemed to have an idea of how much her sons and daughters consumed per week, per number of family members, and their consumption patterns.

My daughter works too, on her own, [in the store] they charge her four hundred, three hundred pesos [her daughter has her own family], and my other son, my son being a man who doesn't earn as much [...] being [single], a young man who doesn't make much; sometimes [in the store] they charge him a hundred, sometimes eighty, sometimes a hundred and fifty, not always, only sometimes [...] because he's a young man, in the afternoon after work he buys what he wants [for himself], potato chips, cookies or Coke.

What Ana is explaining here is how much her son and daughter usually spend on groceries, what her bills are likely to come to in the company store for their supplies at any given time. Ana is detailing how she relied on a family history criterion, based on regular consumption and pricing to estimate whether an amount is correct or not. Doing so implies complex mental calculations where she considered not only the total sum of prices but also pondered the feasibility of the amounts in question as related to the families' general purchasing patterns. It also required being able to estimate what a normal amount for a family purchase could be and compare that to what was being charged in the store as a way of assuring that the storekeeper was not over charging for family purchases.

Strategy 4: Calculating ahead of time

While Ana took on a central role in keeping track of daily purchases, when it came to calculating the total of debts acquired during the week, her eldest sons or daughters were in charge of calculating totals the night before payments were due. However, Ana still compared her children's calculations with the storekeeper's when these did not match:

Interviewer: So, how often do you pay the store? Saturdays?

Ana: Yes, Saturday afternoon … we pay the store …

Interviewer: Does anyone do the accounting here before paying or do you let the storekeeper add it all up?

Ana: I didn't always add it up myself, and when I saw that they were trying to rob me, better to do it too. I would say to one or another [of her children], here, add up the debt, and the other day I went, they added it up: eight hundred … eight hundred seventy, seventy-three, and the other guy was trying to charge me one thousand and some, like one thousand twenty …

Interviewer: That's too much …

Ana: Yes, like one thousand twenty, I said, but here, I tell him, I had someone add it up, I say, how can you charge me more? Ah! He says, eight hundred three, let's see, he's charging more!

Aside from this family strategy of adding up expenses ahead of time, Ana uses *verification criteria based on the trust in her children*, which allows her to stand up to the storekeeper to defend her family's accounting.

These four strategies, together with their underlying criteria, are indicative of the knowledge and diversity of resources that Ana and other non-literate adults have appropriated to help them interact with the storekeeper's annotations and calculations. However, these strategies force them to rely on memory and to consult others who are conventionally literate, which makes these strategies costly in terms of time and effort. However, while relying on memory is not socially verifiable, when more literate others can conventionally write their calculations, their additions become a valid, revisable record. When Ana stood firmly before the storekeeper to confront him for charging excessive amounts, the reasons she would give him were not based on her memory-based strategies but on 'what her children said'. This reliance on others who can read and write numbers contrasts highly with the way Ana described herself as a vendor in a market back in her hometown. When she worked in the market in her hometown, her transactions and interactions with others (clients and vendors) did not require her to interpret or produce written numbers. Ana commented that there she was very skilled at adding up bills 'by memory' and could even do them quicker than others who used pencil and paper or even a calculator:

> And before, earlier, I was a bit better and my head was able to add things up just with memory, just memory, and quickly I would finish before others who used pencil, yes; sometimes they even wanted to charge more than they should [...] I would go to Acapulco to buy things and they added up the bill with calculator or in writing, because I'm buying things to sell, then they want to steal about a hundred pesos or twenty pesos, and I added it all up quick, quick.

This illustrates the importance of mediators in this family's actions to defend themselves against exploitive practices. Furthermore, collaborating with others is not only a strategy used by non-literate adults, it is a common practice in multiple literacy and numeracy events such as paying taxes, filing legal briefs or interpreting medical information (Kalman 1996, 1999, 2005). For workers who are conventionally literate are not exempt from being cheated by storekeepers. As mentioned before, the store's high prices and credit systems leave working families in a highly vulnerable position. From this point of view, checking accounts with others and enlisting their participation is not so much a sign of dependence as it is a healthy strategy for demanding fair treatment in a hostile environment.

Thinking about writing and calculating in contexts of inequality and social tension

This chapter aims to show how the conditions in agricultural camps shape the way that numerical records and calculations are produced. At the same time, it is important to reveal how numerical records and calculations are also part of the social and economic context of workers and employers and how they are constructed in their everyday interactions.

The degree of extreme inequality in agricultural labor is evident throughout Ana's examples and discussion. It is seen across temporary migrant labor camps, where working conditions are unfair and unstable: salaries that have been reduced to the bare minimum; unequal access to formal education and socially valued knowledge, and in asymmetric power relationships marked by the systemic exploitation of the workers and their distrust for their bosses and the company employees.

Numerical practices are not considered 'neutral', they are tainted and shaped by interests and tensions, and here, conflict thrives in these conditions of inequality. Such is the case of the multiple activities that are part of tracking and calculating debts in the company store, where storekeepers wield power through numerical writing and calculations (they are the ones who establish the prices, who write and tally the bills). These conditions strongly influence family strategies for regulating debt and confronting the storekeeper's charges.

The most evident numerical practices—and the most 'officially' recognized—are those manifested by storekeepers. While family calculations happen in more private settings (in their rooms) and on the night before paying their debts, in this time and space, families prepare to face possible differences and conflicts that may arise with storekeepers. Constant tension and lack of trust between those who participate in these activities have led to documenting debts (in clients' and storekeepers' personal notebooks), which can also lead to tensions and possible conflict, even when the storekeeper is in charge of writing in both documents.

Under these conditions, families have developed strategies to prevent and confront potential abuse. While the strategies presented here were employed by an adult woman who cannot read or write, some of these strategies (such as following consumption practices to estimate expenses) are also present in literate adults suggesting how conditions of poverty are infused in numerical practices and can shape knowledge and/or skills related to reading, writing, and numerical calculations.

Families can find a level of certainty by employing strategies supported by verification criteria and by controlling the process of verification and its outcomes. Trust is one of the most predominant criteria: family members who have had a higher level of schooling are given authority to verify accounts; even when others have greater mental calculation abilities and rely on their memory, adults who have been to school take an authority role. In other spaces and in other conditions, the impact of school might be smaller regarding arithmetical practices that do not demand numerical writing and that do not create such widely asymmetrical relationships between participants (such as Ana's interaction with the vendors in her hometown).

Arithmetical practices that emerge because of certain conditions also contribute to how a context is construed. The way these practices take place demonstrates the weakness of those definitions that conceptualize calculating as a neutral skill which individuals can obtain through repetition and mechanical exercise. Here we clearly illustrate how bills are added up by several participants and through multiple events, and how knowledge and know-how are distributed.

Notes

1 Blommaert (2005, 66) notes:
   > That certain discourse forms only become visible and accessible at particular times and under particular conditions is in itself an important phenomenon, which tells us a lot about our societies and ourselves and which necessarily situates particular discourses in the wider sociopolitical environment in which they occur. The stories have a particular 'load' which relates to (and indexes) their place in a particular social, political, and historical moment. Removing this load from the narratives could involve the risk of obscuring the reasons for

their production as well as the fact that they are tied to identifiable people and to particular, uniquely meaningful circumstances that occasioned them.

2 'A high proportion [of adult migrant labor] are unschooled. While on a national level 5.6 percent of men and 8.1 percent of women are illiterate, these numbers are higher among agricultural workers, 18 and 20 percent, respectively' (Barrón 2012). [Translation by authors]

3 Because this chapter was drafted in Spanish, translating the notion of literacy revealed some of its underlying conceptual underpinnings because the word 'literacy' does not exist in Spanish and cannot be reduced to simply 'knowing how to read and write' without considering the social and cultural conditions that are involved in this knowledge. In Spanish, the word *alfabetización* is used under the assumption that someone who is *alfabetizado* 'uses written language to participate in the social world. Therefore, it implies learning how to deliberately and intentionally manipulate written language [...] to participate in culturally valued events and to relate to others' (Kalman 2004, 27; also see Kalman 2008; Kalman and Street 2013). A similar discussion relates to the term *numeracy*; some translate it as *numerosidad* while others use the English word as an attempt to not limit its meaning. In the Spanish version of this chapter, we chose the term *prácticas matemáticas* to underscore the social, historical, and situated nature of human activities that imply mathematical knowledge. We included numerical activities within mathematical activities, and used them to refer to situations where oral and written numbers, as well as mental and written calculations are employed. In this chapter, we will focus on numerical activities.

4 Cole (1985) identifies certain coincidences between psychological theories that use the concept of 'activity' and anthropological perspectives that use the concept of 'event'; in later writing, this author noted that the way both concepts are used and interpreted often makes them appear as synonyms (Cole 1998).

5 In her study, Solares (2012) uses the anthropological theory of didactics (Chevallard, Bosch and Gascón 1998) to identify 'techniques' (ways of solving certain types of tasks) and 'technologies' (discourses regarding these techniques) in agricultural activities that imply writing and numerical calculation. Identifying these techniques and discourses became more feasible when related to agricultural labor activities, but not so much when it came to family economic activities such as paying bills, making purchases, paying for transportation and so on. In this chapter, we used the terms 'strategies' and 'verification criteria' to analyze the ways families took on paying their debts at the stores in the field.

6 About transcription codes in this chapter: [...] omitted/abbreviated data. Conventional punctuation was used to indicate tone and intention (questions, astonishment ...). All names are pseudonyms.

7 Ana's speech was kept as intact as possible; however, we remind the reader that Ana is a speaker of Nahuatl, and, on top of this, we are presenting an edited excerpt of a translation from Spanish to English.

References

Barrón, A. (2012) Dónde están y cómo están los jornaleros agrícolas [Where the agricultural day workers are and how they are doing] *La Jornada del Campo* http://www.jornada.unam.mx/2012/03/17/cam-agricolas.html (accessed 7 November 2016).

Barton, D. and Hamilton, M. (1998) *Local literacies: Reading and writing in one community* London: Routledge.

Blommaert, J. (2005) *Discourse: A critical introduction* Cambridge: Cambridge University Press.

Blommaert, J. (2008) *Grassroots literacy: Writing, identity, and voice in Central Africa* London: Routledge.

Boyarin, J. (1992) *The ethnography of reading* Berkeley, CA: University of California Press.

Broitman, C. (2012) Adultos que inician la escolaridad: sus conocimientos aritméticos y la relación que establecen con el saber y con las matemáticas Unpublished PhD thesis Universidad Nacional de La Plata Facultad de Humanidades y Ciencias de la Educación, Buenos Aires, Argentina.

Canieso-Doronila, M. L. (1996) *Landscapes of literacy: An ethnographic study of functional literacy in marginal Philippine communities* Hamburg: UNESCO.

Carraher, T.N., Carraher, D.W. and Schliemann, A.D. (1985) Mathematics in the streets and in schools. *British Journal of Developmental Psychology*, 3(1): 21–29..

Castanheira, M. (2013) Indexical signs within local and global contexts: Case studies of changes in literacy practices across generations of working class families in Brazil. In Kalman, J., and Street, B. V. (eds), *Literacy and numeracy in Latin America: Local perspectives and beyond* New York: Routledge.

Chevallard, Y., Bosch, M. and Gascón, J. (1998) *Estudiar matemáticas: El eslabón perdido entre enseñanza y aprendizaje.* [*Studying mathematics: The missing link between teaching and learning*] Mexico City: Biblioteca para la Actualización del Magisterio, Secretaría de Educación Pública (SEP),.

Cole, M. (1985) The zone of proximal development: where culture and cognition create each other. In Wertsch, J. (ed), *Culture, communication, and cognition: Vygotskian perspectives* New York: Cambridge University Press.

Cole, M. (1998) *Cultural psychology* Cambridge, MA: Harvard University Press.

Collins, J. and Blot, R. (2003) *Literacy and literacies: Texts, power, and identity* New York: Cambridge University Press.

Cope, B. and Kalantzis, M. (2000) *Multiliteracies: Literacy learning and the design of social futures* London: Routledge.

De Agüero, M. (2006) *El pensamiento práctico de una cuadrilla de pintores: Estrategias para la solución de problemas en situaciones matematizables de la vida cotidiana* [*The practical thinking of a painting crew: Strategies for solving mathematical situations in daily life*] Mexico City: CREFAL – Universidad Iberoamericana.

Delprato, M. F. and Fuenlabrada, I. (2012) *El poder de 'las cuentas': Poder con las cuentas y las cuentas del poder – Problemas de cálculo en la comercialización y preocupaciones sociales de una líder indígena* [*The power of 'accounts': The power of accounting and the accounting of power – Calculation problems in commercialization and social concerns of an indigenous community leader*] Serie Editorial Coloquio. Pátzcuaro, Michoacán, Mexico: Centro de Cooperación Regional para la Educación de Adultos en América Latina y el Caribe.

Dyson, A. H. (1997) *Writing superheroes: Contemporary childhood, popular culture, and classroom literacy* New York: Teachers College Press.

Heath, S. B. (1983) *Ways with words: Language, life, and work in communities and classrooms* Cambridge: Cambridge University Press.

Kalman, J. (1996) Joint composition: The collaborative letter writing of a scribe and his client in Mexico. *Written Communication*, 13(2): 190–220.

Kalman, J. (1999) *Writing on the plaza: The mediated literacy practices of scribes and their clients in Mexico City* Cresskill, NJ: Hampton Press. [Spanish edition, 2003, *Escribir en la plaza*. Victoria Ana Schusseim (trans). Mexico City: Fondo de Cultura Económica.]

Kalman, J. (2001) Everyday paperwork: Literacy and practices in the daily life of unschooled and underschooled women in a semiurban community of Mexico City *Linguistics and Education*, 12(4): 307–391.

Kalman, J. (2002) La importancia del contexto en la alfabetización [The importance of context in literacy] *Revista Interamericana de Educación de Adultos*, 24(3): 11–28.

Kalman, J. (2004) A Bakhtinian perspective on learning to read and write late in life. In Ball, A. and Freedman, S. (eds), *Bakhtinian perspectives on language literacy and learning* Cambridge: Cambridge University Press.

Kalman, J. (2005) Mothers to daughters, pueblo to ciudad: Women´s identity shifts in the construction of a literate self. In Rogers, A. (ed), *Urban literacy. Communication, identity, and learning in development contexts.* Hamburg: UNESCO Institute of Education: Hamburg.

Kalman, J. (2008) Beyond definition: Central concepts for understanding literacy. *International Review of Education*, 54: 523–538.

Kalman, J. (2016) La complejidad de traducir *literacy* [The complexities of translating *literacy*]. In Knobel, M. and Kalman, J. (eds), *Aprendizaje docente y nuevas prácticas de lenguaje: Posibilidades de formación en el giro digital.* [*New literacies and teacher learning: Professional development and the digital turn*] Mexico City: SM Editores,.

Kalman, J. and Street, B. V. (eds) (2009) *Lectura, escritura y matemáticas como prácticas sociales: Diálogos con América Latina* [*Reading, writing, and mathematics as social practice: Dialogues in Latin America*] Mexico City: Siglo XXI Editores / Crefal.

Kalman, J and Street, B. V. (2013) *Literacy and numeracy in Latin America: Local perspectives and beyond* New York: Routledge.

Knijnik, G. (2003) Educación de personas adultas y etnomatemáticas. Reflexiones desde la ucha del Movimiento sin Tierra de Brasil [Adult education and ethnomathematics. Reflections from Brazil's Sin Tierra Movement]. *Revista Decisio* 4: 8–11.

Levine, K. (1986) *The social context of literacy* New York: Taylor & Francis.

Lave, J. (1988). *Cognition in practice: Mind, mathematics and culture in everyday life.* Cambridge: Cambridge University Press.

Lave, J. (2011) *Apprenticeship in critical ethnographic practice* Chicago, IL: University of Chicago Press.

Maldonado, S. (2005) Denuncian explotación de jóvenes contratados para pizcar uva en Sonora [Exploitation of young people contracted to pick grapes in Sonora denounced]. http://www.jornada.unam.mx/2005/05/24/index.php?section=estadosa ndarticle=041n1est (accessed 7 November 2016).

Moreo Gastelum, J. (2015) Seguridad Alimenticia MX. http://seguridadalimentariamxson. blogspot.mx/ (accessed 20 January 2018).

Muñoz A. (2012) Cómo se transportan l@s jornaler@s migrantes [How farmworkers are transported]. *La Jornada del Campo.* https://issuu.com/la_jornada_del_campo/docs/ la_jornada_del_campo_54 (accessed 7 November 2016).

Moss, B. J. (1994) *Literacy across communities* Creskill, NJ: Hampton Press.

Padilla, E. (2015) Conocimientos matemáticos de menores trabajadores: El caso de la proporcionalidad. Unpublished Masters thesis. Universidad Pedagógica Nacional, Mexico City.

Reder S. (1994) Practice-engagement theory: A sociocultural approach to literacy across languages and cultures. In Ferdman, B., Weber, R. M. and Ramirez, A. (eds), *Literacy across languages and cultures* Albany, NY: State University of New York Press.

Reder, S. and Dávila, E. (2005) Context and literacy practices. *Annual Review of Applied Linguistics*, 25: 170–187.

Scribner, S. and Cole, M. (1980) *The psychology of literacy.* Cambridge: Cambridge University Press.

Solares, D. (2012) Conocimientos matemáticos de niños y niñas jornaleros agrícolas migrantes [The mathematical knowledge of child farm laborers]. Unpublished PhD thesis. Departamento de Investigaciones Educativas, Centro de Investigaciones y Estudios Avanzados, México.

Soto, I. (2001) Aportaciones a la discusión sobre la enseñanza de las matemáticas a partir de la didáctica y la etnomatemática [Contributions to the discussion of how to teach math and ethnomathematics]. In Lizarzaburu, A. and Zapata, G. (eds) *Pluriculturalidad y aprendizaje de la matemática en América Latina: Experiencias y desafíos. [Pluriculturalism and teaching mathematics in Latin America].* Madrid: Ediciones Morata.

Street, B. V. (1984) *Literacy in theory and practice* Cambridge: Cambridge University Press.

Street, B. V. (1995) *Social literacies: Critical approaches to literacy in development, ethnography, and education* London: Longman.

Tett, L., Hamilton, M. and Hillier, Y. (2006) *Adult literacy, numeracy, and language: Policy, practice and research. Maidenhead:* McGraw-Hill Education.

Valsiner, J. and Rosa, A. (2007) Introduction: Contemporary socio-cultural research uniting culture, society, and psychology. In Rosa, A and Valsiner, J. (eds) *The Cambridge handbook of sociocultural psychology* Cambridge: Cambridge University Press.

Wagner, D. A. (1993) *Literacy, culture, and development: Becoming literate in Morocco* Cambridge: Cambridge University Press.

Wogan, P. (2004) *Magical writing in Salasaca: Literacy and power in highland Ecuador* Boulder, CO: Westview Press.

Zavala, V. (2002) *Desencuentros con la escritura: Escuela y comunidad en los Andes peruanos [Falling out with writing: School and community in the Peruvian Andes]* Lima: Red para el desarrollo de las ciencias sociales en Perú

5

MATHEMATICS IN PRE-VOCATIONAL EDUCATION

A model for interfaces between two different teaching contents

Lisa Björklund Boistrup, Elisabet Bellander and Michael Blaesild

Introduction: Mathematics, pre-vocational education and working life

This chapter adds to research where the interest is in the role of mathematics in workplace activities and vocational/professional education (see, for example, Bakker and FitzSimons 2014). Studies in this field have been performed in relation to vocations requiring relatively little post-school education (e.g., Keogh, Maguire and O'Donoghue 2016), as well as to professions requiring longer education (e.g., Frejd and Bergsten 2016). The vocation which is the focus of this chapter is construction work, and the specific interest is on mathematics as part of vocational students' anticipated work. On the one hand we present an empirical project where teachers work in a collaborative teaching approach, together with students and a researcher, providing insights into how the significance of mathematical activities in construction work can be made more visible, recognised, and demarcated. On the other hand, our findings are part of theory development in that we present a model for how interfaces (see Damianian and Sträßer 2009) between mathematics and construction work, as teaching content, can be understood.[1] The pre-vocational study programmes (from here on called vocational education) in Sweden, where the study was performed, are part of the upper secondary school system. We view mathematics as a human activity, and include here all such activities throughout history which have come to be labelled as mathematics (FitzSimons 2002). As a consequence, we adopt a broad understanding of what constitutes mathematics.

During mathematics teaching, if the vocation is not represented in an authentic way, the relationships between mathematics and real workplace practice are not made clear to the students (FitzSimons 2014b, see also Boistrup and Keogh

2017). Hence, in order for mathematics as part of workplace activities to be, or become, relevant content within the framework of vocational education, much more is required than a superficial understanding of the vocation Boistrup and Gustafsson 2014; Wedege 1999; Williams and Wake 2007). This is supported by studies on mathematics in Swedish vocational programmes in general (Lindberg and Grevholm 2011; Muhrman 2016). Related to this theme, Gellert and Jablonka (2009) observe that when examples from everyday context are announced in classroom communication, they can only be "almost real", since the communicational context is located within a mathematics class. This may be described in terms of life outside the classroom as being looked upon with a (school) mathematical "gaze" (Dowling 1998). The adoption of this gaze can, however, differ between students from different social groups, where, for example, middle-class students may do this more easily than working-class students (Gellert and Jablonka 2009), and consequently can more easily understand what counts as legitimate knowledge within the school setting.

A vocational training programme in Sweden takes three years to complete. It includes education in the vocational teaching subject, such as construction work, and school subjects such as mathematics, English, and Swedish, as well as other teaching subjects. For mathematics teachers and vocational teachers in such vocational programmes, there is a requirement within the system that the mathematics teaching should to a large extent be connected to the respective vocational education (Skolverket 2011), even where the mathematical content is expected to vary between vocational programmes (Dahl 2014). As authors of this chapter, we want to emphasise that we view mathematics as a discipline with its own relevant content in vocational programmes alongside mathematics which is an integrated part of workplace practices (see also Wheelahan 2009). Dahl (2014) describes this in terms of unequal possibilities for vocational students to reach the main stated goal of the schooling system of Sweden, which is that all students should receive a solid grounding for personal development and for active participation in society. The focus of this chapter is, however, primarily on mathematics related to vocational teaching content, namely construction work, but we address to some extent content that is mainly labelled as school mathematics. The purpose of the research project was to deepen the knowledge about how interfaces between mathematics and construction work, as teaching content, can be characterised. The following research questions specified the study:

1   How can interfaces between the teaching contents of mathematics and construction work be captured in a theoretically and empirically derived model?
2   How do students, teachers and a researcher characterise interfaces between mathematics and construction work in answers to written questions?

The purpose and research questions above specifically address both school mathematics and the teaching subject of construction work. In the discussion we point to general implications of the model in terms of relationships between mathematics and other vocational teaching contents.

The contextual background of the study

The contextual background of this study is a project initiated by two authors of this chapter (Elisabet and Michael). The purpose of this professional development project was to teach together, and to find points of contact between the teaching subjects of mathematics (Elisabet) and construction work (Michael) in a way that engaged teachers as well as students. Before the actual teaching, a plan was made with specific attention to the collaboration in order for the overall teaching and learning to benefit as much as possible from both subjects. The purpose was also to demonstrate clearly to the students how the subjects were interconnected. In practice the procedure was that Elisabet, as a mathematics teacher, visited Michael's vocational upper secondary school each week and accompanied him in his teaching as a construction work teacher. The regular mathematics teaching in the mathematics classroom was conducted by other mathematics teachers. The lessons in which Elisabet participated were partly those where the students were performing real construction work in an outdoor area of the school.

When the project began, it was viewed as essential for the students to experience the two school subjects as a whole instead of as separated in the traditional manner. During the joint teaching, questions arose about the characteristics of the teaching resulting from this professional development project. In discussions with other teachers, and also decision-makers on a municipality level, Elisabet and Michael experienced the need for an in-depth elaboration on how the relationships between mathematics and construction work, in the form of collaborative teaching, may be described. This led to the initiation of the research study described in this chapter.

Theories for understanding relationships between mathematics and vocational practices.

As a theoretical framework, we mainly adopted a selection of concepts by Bernstein (2000). Bernstein and his colleagues worked empirically and theoretically to construct models showing different aspects of education systems, mainly with an interest in how and why some student groups succeed in school, and others do not. Bernstein also had an interest in teaching contexts in a broad sense, and included, for example, workplaces (see e.g., FitzSimons and Boistrup 2017). In circumstances that may be counted as informal education contexts, there is a:

> purposeful intention to initiate, modify, develop or change knowledge, conduct or practice by someone or something which already possesses,

or has access to, the necessary resources *and* the means of evaluating the acquisition.

(Bernstein 2000, 199–200)

Even though our investigation was conducted within the formal context of upper secondary schooling in Sweden, this quotation by Bernstein is relevant, since we examined school mathematics in relation to authentic workplace activities as educational content (described in the section on methodology below). This means that in this chapter we do not discriminate between tasks as part of vocational education and tasks as part of real working life (cf. Lindberg 2003).

Recontextualisation and pedagogic discourses

The main theoretical concept of Bernstein (2000) adopted in this study is *recontextualisation*, and we describe this concept here with other related concepts of Bernstein (2000), while drawing also on FitzSimons (2014b) and FitzSimons and Boistrup (2017). His claim is that there are principles underlying all 'pedagogisation' that takes place in society. These principles address how different knowledge is transformed between contexts. Connected to these principles are rules operating at three levels in society, macro-level (at the political steering level), meso-level (e.g., at the school level), and micro-level (at the classroom level through interactions between students and teachers).

On the macro-level are the *distributive rules*. Through these rules, different knowledge is distributed, as well as different forms of knowledge, to different groups of people (see Dahl 2014; see also Wheelahan 2009). On the meso-level, the *recontextualising rules* regulate how the pedagogic discourse is formed, for example in school. Bernstein states that the principles of recontextualisation steer towards a selection of knowledge, which "appropriates, relocates, refocuses and relates other discourses to constitute its own order" (Bernstein 2000, 33). One example is how the real workplace discourse of carpentry in the world of schooling is transformed into the imaginary school discourse of woodwork. This is also what happens when teachers in vocational school subjects recontextualise vocational disciplines to vocational teaching content. Finally, on the micro-level, there are *evaluative rules*, which form the pedagogic practice (e.g., in school or at work) in terms of how the individual acquires the expected knowledge and how this is evaluated.

Recontextualisation, when presented in mathematics education research, often concerns how mathematics teachers transform the discipline of mathematics to their teaching in the form of school mathematics (Lerman 2000). FitzSimons (2014b; see also FitzSimons and Boistrup 2017) emphasises that recontextualisation of mathematics is not only relevant for mathematics teachers, but for all workers, in the sense of the transformation of one's mathematical knowing into the day-to-day demands of working life. This is the perspective from which we adopt the term recontextualisation in this chapter.

Bernstein (2000) also described what he called *vertical* and *horizontal* discourses. These are two different forms of knowledge structures and they constitute a way to describe the "what-question" in education systems. In this chapter, when we write about recontextualisation, it is in relation to aspects of these two discourses. *Vertical discourse* is about knowledge in a discipline, for example academic mathematics, and is described as theoretical, concept-based and with a generalisable knowledge (Bernstein 2000). Knowledge from this discourse may be "made" specific to a context through recontextualisation, which happens for example when a mathematics teacher makes theoretical mathematics concepts possible for the students to understand through the use of a variety of strategies, artefacts, etc. When a person has understood mathematical concepts and integrated them into their repertoire of, for example, vocational knowledge, concepts of this kind may be used in different kinds of activities, within a certain vocation and in other contexts. This may be compared with an everyday skill, which is easiest learned in the context where it normally is used, but may not be useful in other contexts. The discourse of knowledge corresponding to this skill is labelled the *horizontal discourse* by Bernstein (2000).

Within the vertical discourse, there are scientific disciplines with a single hierarchical structure of knowledge. Mathematics is classified by Bernstein as a discipline within the vertical discourse, but with a horizontal knowledge structure (see also FitzSimons and Boistrup 2017). The reason is that mathematics has several parallel languages, for example algebra and geometry. Within the vertical discourse there are also professions with clear knowledge criteria, such as part of trades. Vocational knowledge within a trade or craft, such as electricians' knowledge, is counted as a vertical knowledge discourse (with an implicit transmission of knowledge through apprenticeship), while the knowledge connected to each specific workplace is an example of a horizontal knowledge structure. In our analysis, we have mainly operationalised the concept of recontextualisation, but we have also connected this to vertical and horizontal discourses.

Methodology

The project was carried out as a participatory action research project where the participants were teachers (Elisabet and Michael), a researcher (Lisa), and students (10). An action research project concerns people within a system who want to develop something within the system, and simultaneously gain knowledge about this "something". Consequently, action research constitutes meeting points between professional development and research, where both these parties are active throughout the project. An action research project runs in cycles where analysis of the current situation is made, with subsequent changes, followed by a new analysis and new reflections, and so on (Carr and Kemmis 2003).

This kind of action research is social in two different ways (Atweh 2005). One is the viewing of mathematics teaching and learning as part of a larger context where there are limitations, decision makings, and planning. The teaching of mathematics is normally undertaken within the institution of schooling and this affects in different ways what is taking place.

Participatory action research is also social in the sense that the participants problematise the research itself. Hence, power relations between participants are paid critical attention. In the research there is an aim to make use of the participants' different experiences, and to listen to all people in the project. In our project we also took these social aspects into account through the theories we adopted, where Bernstein's sociological theories address human interaction as a part of, and affected by, the broader institutional context. We also strove towards a situation where the students were not only participants in the teaching, but in the research as such.

The study was carried out during the first half of 2016. The focus of the project was on the part played by the teaching where the students performed real construction work in an outdoor area. The students planned together with the teachers how they wanted to equip the area with pathways, stairs, a pond, etc. These constructions were not made (only) in the sense of students' practices, but it was actually a real project where the constructions were going to be kept as part of a recreational area of the school. This means that there were not any artificial activities taking place, in the way found in other studies on vocational education (Lindberg 2003), although Michael, as construction work teacher, taught the specifics of the subject content in connection to the construction project. Elisabet, as mathematics teacher, followed the construction work and discussed the role of mathematics with students while the work was going on. She also, sometimes, addressed specific mathematics content, while drawing on previous or anticipated construction work activities.

Teachers and researcher met regularly, and likewise students and teachers (in connection to the teachers' common teaching). The whole researching group (teachers, researcher and students) met over four days altogether.

The data analysed in this chapter consists of the following:

- Videos from the teaching where the teachers in the project cooperated with a focus on points of contact between mathematics and the students' anticipated workplace practices.
- Photographs from teaching and from meetings between teachers and researcher.
- Logs written by teachers and researcher, and notes from meetings.
- Written answers to four "participant questions" posed in a questionnaire to all participants (teachers, researcher and students), at the beginning and the end of the project. The participant questions were distributed on two pages as follows:
  - Page 1: When is the mathematics part of the teaching of construction work? How do you perceive this?
  - Page 2: When do you think that construction work is part of the mathematics classroom? How do you perceive this?

We adopted somewhat different analytical strategies for the two overarching research questions. For the analysis of research question 1, on developing a model for interfaces between mathematics and construction work, we went between theoretical assumptions and our attempts to sketch tentative models. Additionally, the process here was an interplay between the writing of two texts. One was a theoretical text about the "hiatus" between mathematics and vocational contexts, and how this may possibly be overcome. This text is written by FitzSimons and Boistrup (2017), and was written partly during the same time period as the analytical processing of the action research project described in this chapter took place. Lisa in the role of co-author in both texts could then bring insights from one text to the other, and vice versa. In this way, the theoretical framework in this chapter is affected by both Bernstein (2000) and how some of Bernstein's concepts have been adopted in FitzSimons and Boistrup (2017; see also FitzSimons 2014a, 2014b). Likewise, the discussion of implications in FitzSimons and Boistrup were affected by the action research described in this chapter.

For the analysis of research question 2, on how students, teachers and researcher characterise interfaces between mathematics and construction work, we began by acquainting ourselves with the data, and we tentatively distinguished themes related to the model in research question 1. This way Bernstein's concept of recontextualisation also affected the analysis for research question 2.

Analysis and findings 1: a model for interfaces between the teaching content of mathematics and construction work

Our analytical process when developing a model for interfaces between the teaching content of mathematics and construction work was characterised by exchanges between inspiration from theoretical constructions and examinations of the empirical data. The final product was a model based both on some of Bernstein's theoretical concepts, and on our specific research undertaken together with the students. Below we give an account of the analytical process and of the final model.

Earlier versions of the model

An initial purpose of the research project was to find ways of describing relationships between mathematics and construction work. The diagram according to which we worked at the beginning of the project can be shown as Figure 5.1.

Figure 5.1 shows how we initially saw construction work as 'another' content area than school mathematics, and how we focused on finding the 'connections' between the areas, as shown by the two arrows. We did not explicitly draw this

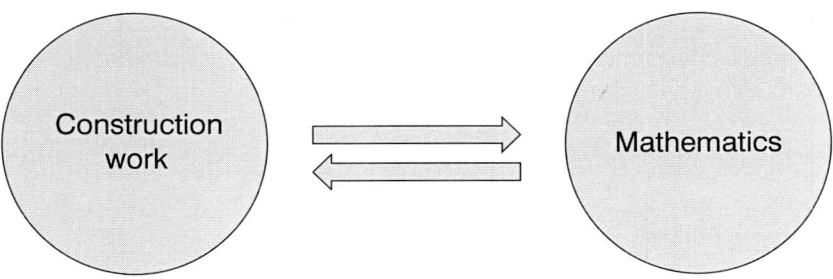

FIGURE 5.1 Digram illustrating how we viewed the relation between the two content areas at the beginning of the project

picture in this phase of the project, but with gestures and speech it was a model of this kind that we implicitly were referring to when discussing with each other (teachers, students, researcher). Relationships that we envisioned could, for example, be the construction work teacher paying attention to mathematics as shown through calculations. It could also be the mathematics teacher using examples from construction work when teaching mathematics.

A first step towards a more nuanced model was taken after the first collection of written answers to the participant questions. When we transferred the handwritten data to a computer and started to discuss tentative themes, we also discussed ways of describing relationships between mathematics and construction work. After a while, we produced a first sketch of a model (Figure 5.2).

When we drew the tentative model in Figure 5.2, we were discussing in what contexts we could identify mathematics. We reasoned about mathematics (upper circle) and how it may be related to different vocations (V in the figure), and that in this case mathematics is part of a specific context. At the same time, it is possible to see mathematics as one out of several aspects of a vocation (circle in the middle). Furthermore, there is also mathematics (bottom circle) which is more connected to educational or research contexts than to everyday contexts. The arrows in Figure 5.2 display different possible relationships between mathematics and vocational contexts.

In the next step we, teachers and researcher, adopted a view of mathematics as *recontextualised* in vocational practices, along with the discussion of *vertical* and *horizontal* knowledge discourses according to Bernstein (2000; see also FitzSimons 2014b). FitzSimons emphasises that recontextualisation of mathematics in vocational practices should be an essential teaching content per se in vocational education. In this phase of the project we started to discuss whether there exists a space of interfaces between mathematics and construction work which can be viewed as a teaching content area on its own, and in which mathematics can be relocated, refocused, and related to vocational activities. This would then be an area which is neither first and foremost mathematics nor construction work, but

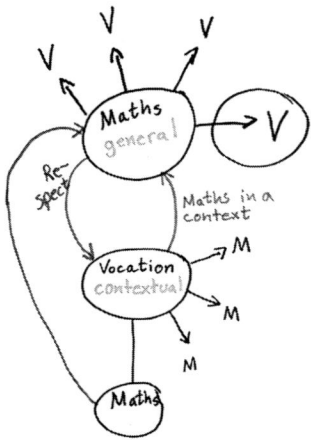

FIGURE 5.2   Sketch of a possible model of interfaces between the two content areas

rather both. During these discussions we also used videos and photos from the teaching activities.

One consequence of our discussions on the data in relation to the theoretical concepts was that we wanted our model to illustrate the interfaces between mathematics and construction work as a teaching and learning area on its own. In parallel, we investigated what kind of recontextualisations of mathematics may take place in construction work. With inspiration from research on how mathematics is adopted in professions requiring lengthy periods of mathematical education (for example, Frejd and Bergsten 2016), we could also read explicit use of mathematics into our data. One example was when the students were involved in authentic construction work, and they explicitly used mathematics to plan that work. In parallel we could read mathematics as an integrated part of vocational activities, in the same way as described by Johansson (2014), where mathematics as part of the work by a nursing aide was analysed. The following excerpt is taken from the occasion when a preliminary version of the model in Figure 5.3 was developed.

> To engage in foundational mathematics is preparing for possibilities to also engage in more difficult mathematics such as trigonometry, where also another type of recontextualisation may occur.
>
> We went through Lisa's document about vertical and horizontal structures and discussed the teaching in our project outgoing from that.
>
> Recontextualisation as a teaching content on its own. A responsibility for both mathematics teachers and vocational teachers. And here is, really, two types of recontextualisation, A and B in Lisa's document.
>
> The importance of

- showing respect for each other's teaching content;
- not only looking for the visible mathematics;
- discussing with the students that there are two types of connections;
- the importance of [sic] that also the B-version of recontextualisation needs mathematical knowledge

(Translated from notes made by one of the teachers during the second half of the action research project)

In this excerpt it is clear how our, both teachers' and researcher's, interpretations of Bernstein's concepts of recontextualisation, and also of vertical and horizontal discourse, were starting points for the development of a model. The text in "Lisa's document" referred to above concerning vertical and horizontal structures was copied from a previous version of the article by FitzSimons and Boistrup (2017), which explained how vertical discourses also include horizontal knowledge structures. This applies not only to mathematics (with strong grammar, according to Bernstein 2000), but also to trades such as the construction trade (with weak grammar). The fact that both of these knowledge areas belong to what Bernstein denotes as vertical discourses inspired us to try to integrate both knowledge areas equally in the same model.

In the conversations reflected in the excerpt, we also discussed different ways that we could see recontextualisation of mathematics in construction work. We called these A and B in "Lisa's document" and a new kind of model was created during this discussion.

A model for interfaces between mathematics and construction work

In Figure 5.3 the final model is shown, which also represents our findings to the first research question. A modified version of this model, in the form of a table appears in FitzSimons and Boistrup (2017). A first basis for the model was our theoretical framework with Bernstein's concepts. A second basis was previous research on mathematics in working life. A third basis was our analysis of collected data in the form of texts, pictures and videos.

Subset 1 in the model represents activities in construction work where there is no significant recontextualisation of mathematics. This could be the situation when a construction worker is digging and evening out the ground (Figure 5.4).

In the planning for the activity in Figure 5.4 mathematics has probably played a significant role, but when the actual digging is taking place mathematics is not a significant aspect.

Subset 2 of the model is when construction work meets mathematics. As noted above, we identified two types of recontextualisation of mathematics. In one, 2A, a construction worker, or student in the construction programme, uses mathematics explicitly to solve a task. S/he then is not actively involved in the full activity on the actual construction site, but is rather in a separate area or

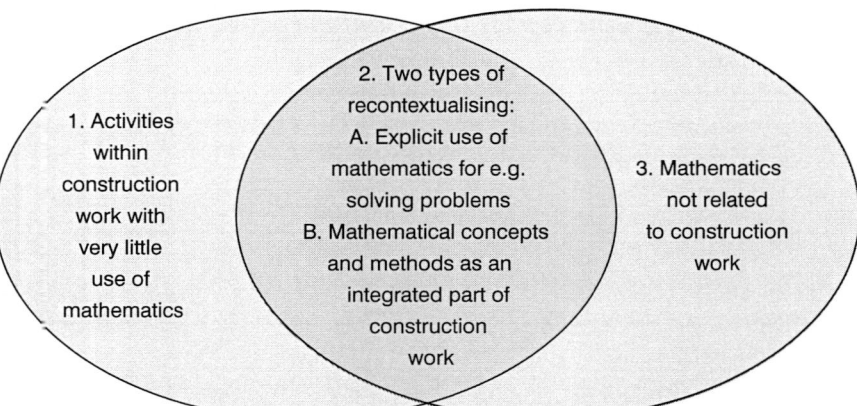

FIGURE 5.3 Model for relationships between construction work and mathematics

in an indoor office. In the planning phase this type of recontextualisation is common. A responsible construction worker has a drawn plan as the starting point for the planning of the practical construction work. In the planning are included calculations concerning the implementation of the plans, material consumption, ordering of material (where a certain loss must be taken into account), and schedules for the team, machines etc. When *relocating* mathematics

FIGURE 5.4 Removing sand (source: photo taken with mobile phone during activities with the students in the project)

from one discourse (mathematics) to another (construction work) there is a constant need to also *refocus* its use: for example, when calculating the number of lorries needed to remove sand and soil from the site. The volume can often easily be calculated through using the formula for volume of a rectangular block (*relocating*). However, when the digging starts, the volume of the mass to be removed increases according to a factor correlated to the specific type of ground. This factor is then necessary to be included in the mathematical model in use (*refocusing*). In similar ways, students in the construction programme may have to plan the real construction work they are expected to perform as part of the construction teaching and learning activity. In Figure 5.5 the planning of an outdoors staircase is shown.

The use of mathematics in line with 2A in Figure 5.3 may also occur when a problem appears during the practical construction work, for example if something breaks, and methods to fix it are needed urgently to keep to the time frame. The problem may then be solved while the student/worker moves away to a quieter area, thinking about the issue. Sometimes the solution to the problem includes the application of a geometric formula.

In the other type of subset 2 in the model, 2B, relocating and refocusing of mathematics is an integrated part of construction work. This is a type of recontextualisation where construction students/workers, often on the construction site, estimate numbers or measurements and the like. Here, mathematics is not always easy to distinguish from the vocational context, even though mathematical errors could compromise the construction project or even safety or security. In the construction work of this study, this could concern estimations of how deep to dig when tiling, or to position tiles with different

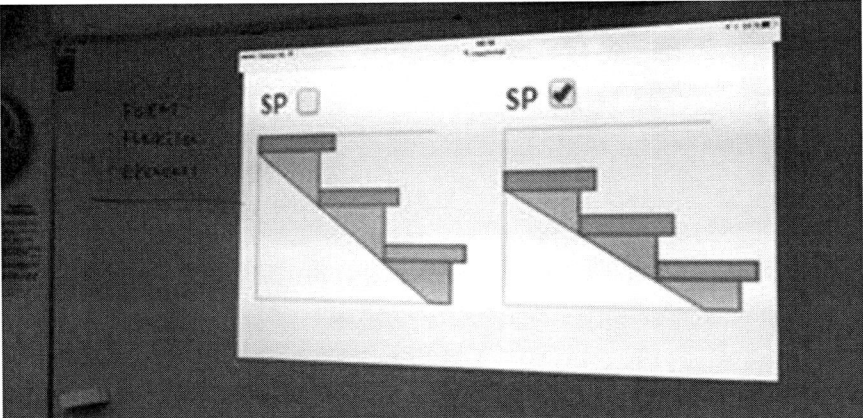

FIGURE 5.5 The students have jointly decided to build an outdoor staircase, which then needs to have a certain inclination. Explicit calculations are needed in order to get the height, length, and width measurements, and also a diagonal with a certain inclination (source: photo taken with mobile phone during activities with the students in the project)

FIGURE 5.6 Use of mathematical concepts and methods to build a staircase (source: photo taken with mobile phone during activities with the students in the project)

measurements including thickness, and to keep an even distance between the tiles so that it evens out through the row of tiles. In this kind of work, a worker makes good use of her/his experiences. Figure 5.6 shows how the staircase that was planned in Figure 5.5 was built as part of the practical work by the students.

In the activity in Figure 5.6, mathematics is integrated as part of other work tasks in line with 2B in Figure 5.3. Subset 3 in the model in Figure 5.3 concerns mathematics teaching content without any connection to a workplace context. Here the context is mainly the mathematical content itself. This may, for example, include students practising the use of a mathematics formula making connections between trigonometry and Pythagoras's theorem.

Analysis and findings 2: student, teacher and researcher voices on relationships between mathematics and construction work

Here we describe our analysis of the written answers to the participant questions (students, teachers and researcher). Our analysis is built on how we interpreted the answers to the participant questions in the context they were written. This means that our interpretations are not built only on the sentences given here as examples, but also on what, for example, a student had written just before or afterwards.

The headings below constitute the different categories that we could discern from the answers to the participant questions as a whole. The point of departure for the analysis was the model in Figure 5.3, where we identified nuances in the model's subsets, and we could also discern relationships *between* the subsets in the model. The headings in the categories correspond to the subsets in the model in Figure 5.3.

1. Construction work without a clear presence of mathematics

There were no clear answers in the data where construction work was emphasised without mathematics. This is clearly due to the fact that the questions addressed mathematics in construction work, and vice versa.

2A. Mathematics is mentioned as explicitly used in construction work

Among the student answers to the participant questions there were examples of mathematics explicitly used in construction work. One theme here was the planning of work tasks in construction projects with mathematics as a clearly identifiable part:

> Calculating the things we are going to build, then we use mathematics.

> It concerns how we calculate materials.

> In outdoor construction work we need to calculate how many tiles that fit into a form.

> When one must calculate an area of for example frames. We did that in the workshop.
>
> *(Different students' answers)*

Also among the answers from the group of teachers/researcher, there were examples of mathematics as part of planning:

> Which machines do we need to have in order to load and carry; what it costs in terms of budget, compared to if we use other materials; what is appropriate – reasonable.

> Mathematics is most clearly present in the planning phase, when the students are going to plan an assignment where they have to calculate dimensions, surfaces, volumes and lengths primarily.
>
> *(Answers from teachers/researcher)*

In the data there was also a presence of explicit use of mathematics as part of solutions of smaller or larger problems, which appear during the construction phase.

> Another kind of situation may be when a situation occurs where some values (measurements) need to be found out for a construction, and where it is important that it is exactly correct. A knowledgeable worker may withdraw her/himself from the practical work and use for example geometrical formulas to calculate the needed value, adopting formula sheets, or a construction reference book. Also here profound knowledge about the construction context is crucial for the measurements to be correct.
>
> *(Answers from teachers/researcher)*

Overall, the excerpts above reflect how the answers to the participant questions contained descriptions of explicit recontextualisation of mathematics in construction work, both as part of planning, and as part of problem solving.

2B. Mathematical concepts and methods are characterised as an integrated part of construction work

Among the answers to the participant questions there were examples of mathematical concepts and methods as an integrated part of construction work:

> When we build something, then we use maths all the time.

> Much measuring.

> It is really fun to both work and calculate.

> Maths is also present in the workshop [outdoors], when you for example dig a certain area.
>
> *(Different students' answers)*

The answers above show that the students sometimes characterised mathematics as integrated in construction work in general terms (for example, "Much measuring") and sometimes more specifically (for example, "when you dig a certain area"). Also among the answers from teachers/researcher it was possible to identify mathematics as an integrated part of construction work:

> Symmetry – positioning stones when building a pond in a specific pattern – mirror symmetry.

> The mathematics that the students use is mainly assigning measurements of lengths, heights, and angles. In this phase the students also have to

make assessments about, for example, where to place the soil dug from somewhere else, and how much still needs to be dug.

(Answers from teachers/researcher)

Overall, the excerpts above reflect how the answers contained descriptions of more implicit recontextualisation of mathematics in construction work. These work tasks could, for example, be building, measuring, digging and tiling.

3. Mathematics without any connection to construction work

Among the answers we could also find the category where school mathematics was described without any clear connection to construction work. These could be short descriptions of what kind of mathematics was relevant for construction work, without an explanation of how:

When we kind of calculate times of length and height.

Geometry, angles, degrees.

It may be addition, subtraction, multiplication, division.

(Different students' answers)

The answers above reflect how different areas of school mathematics were lined up, without a clear connection to construction work. Also among the answers from teachers/researcher this theme was present:

Areas – relationships – enlargement/reduction/plausibility.

Especially that in the construction work school subject is included lots of geometry, numbers, probability, all the areas that are addressed

(Answers from teachers/researcher)

We could see a difference between the first and the second time the participants answered the questions: The answers from the second time more clearly described how mathematics could be recontextualised in the activities of construction work.

Other answers concerned how the regular mathematics teaching did not connect to construction work to a great extent. In this case it was about students describing their experiences:

I do not notice it [construction work in the mathematics classroom] and I do not think that it creates a difference.

You calculate differently in the [mathematics] classroom since there you sit with a textbook and calculate, calculate and calculate.

> When you are in the [mathematics] class you do not understand, and it is rather complicated.
>
> *(Different students' answers)*

What we read from the student statements above is how these students do not see construction work as part of their regular mathematics teaching. Moreover, the textbook does not seem to connect to the vocational content to a high degree. The answers show that some students saw an absence of vocational context in the regular mathematics teaching, and that the mathematics itself was hard to understand. In the following we present some categories which address relationships between subsets of the model in Figure 5.3.

Construction work together with mathematics may explain/ help/affect the learning of mathematics

A category which describes a relationship between subsets of the model in Figure 5.3 is that teaching in construction work together with mathematics teaching may explain and help the learning of mathematics:

> To use mathematics in construction work is better and easier to learn and to understand, and that they [the teachers] show it so that you can see it and get it easier into your head.
>
> Personally I learn better through practical work.
>
> It is easier when [s/he] for example gives an example of how would you do if you had a wall and did it like that. So then it gets better.
>
> It makes it easier to understand mathematics when you for example know why you do it.
>
> Think it is more fun to use mathematics in construction work because then I understand more and you see HOW you should calculate.
>
> *(Different students' answers)*

The answers above show how some students saw a clear advantage in learning mathematics through a teaching activity which was organised as collaboration between mathematics and construction work. They addressed how mathematics became easier to understand and also that the learning was affected so that it became more fun. In the data with the teachers' and researcher's answers, the same theme was possible to identify, but then it was rather, and not surprisingly, expressed from a teaching perspective:

> I can also see that my mathematics language changes since I readily include construction language and relate more to that.

I have discussed relationships between tiles' proportions – uniformity when the students build a pathway in theory.

According to the teachers, the students' attitudes have changed during the course of the project and they are nowadays not stressed when the mathematics teacher wants to discuss work tasks. Instead they stop and discuss and are open to potential connections to mathematics.

(Answers from teachers/researcher)

From the teachers'/researcher's answers it is possible to read a description of how teaching mathematics *and* construction work together could support school mathematics when the language was clearly connected also to the vocational teaching content. Another theme is how the mathematics teacher could capture mathematics as part of construction work tasks, for example of tiles' proportions. The students also characterised their experiences of being more positive when the latter took place.

Mathematics may explain/help/affect the learning of construction work

A relationship between the subsets of the model in Figure 5.3 concerned how a stronger presence of mathematics in the joint teaching could help the learning of construction work. Answers analysed this way were mainly present in the second round of answers:

It [mathematics] explains how things work.

You learn different ways of calculating materials and how far a distance and to dig deep.

It is good fun that you simultaneously work and calculate. You get more fulfilment from work. Not dull and hard.

(Different students' answers)

From answers such as the above, we read how the students described mathematics on one hand as being helpful in understanding certain procedures in construction work, and on the other hand as helpful in calculating certain critical measurements. The third statement shows a student expressing that the presence of mathematics positively affected their construction work.

Also in the group of teachers/researcher, the category that mathematics explained/ helped/ affected construction work was present:

A central theme is the breadth of the school subject construction work, including the many images. We can call it a 'red thread' [a Swedish expression which indicates how a notion is running through a text or a sequence of events]. On this thread we can attach mathematics which may strengthen the content, for example when discussing preparations of the creation of a space for a building project ...

I experience that the core of the vocation is more clearly illuminated, in the way that I get an easier task to explain why we perform constructions in a certain way, and the building materials dimensions.

(Answers from teachers/researcher)

From the excerpts above we read that from the teachers' perspective also, it was possible to see positive consequences when mathematics was given the role to clarify and to explain constructions, for example.

Summary of findings and conclusions

In the following we summarise the findings from both research questions, where, in brief, one was concerned with how interfaces between mathematics and construction work may be captured in a model, and the other was concerned with how students, teachers and a researcher characterise interfaces between mathematics and construction work. In Figure 5.7 we show a reduced version of the model and summarise categories, also in terms of the findings of RQ 2.

In summary, the following subsets, and relationships between them, were identified in the analysis:

- Subset 1. Activities within construction work making very little use of mathematics.
- Subset 2A. Recontextualisation as an explicit use of mathematics (e.g. for solving problems).
- Subset 2B. Recontextualisation of mathematical concepts and procedures as an integrated part of construction work.
- Subset 3. Mathematics without any connection to construction work.
- Construction work together with mathematics (subset 2 in Figure 5.7) may explain/help/affect the learning of mathematics (subset 3).
- Mathematics (subset 3) may explain/help/affect the learning of construction work (subsets 1 and 2).

Discussion

The main contribution from this chapter is the model shedding light on the teaching content of mathematics and construction work within vocational education. Through an analysis of written student answers, and also teacher and researcher answers, we were able to bring the participants' voices into an elaboration of the model. The findings of this chapter should not be taken to

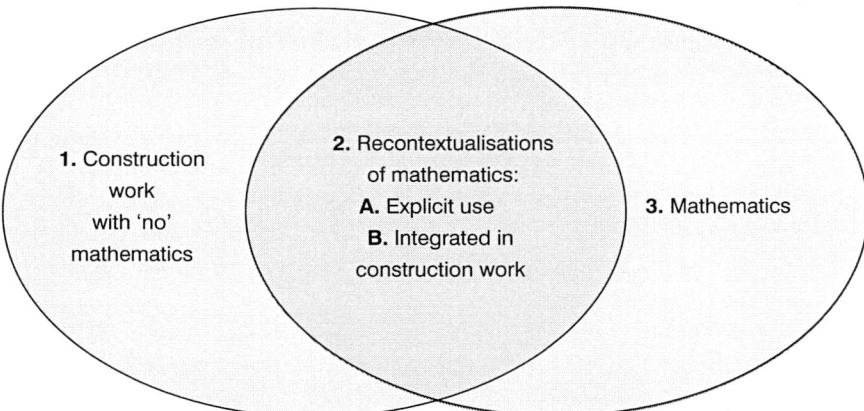

FIGURE 5.7   Reduced version of model for relationships between construction work and mathematics

imply that any cooperation would always benefit students' learning. Something made clear throughout the project was that both content areas, construction work and mathematics, were taken seriously, which happened through experienced teachers' well-informed planning in the context of respectful collaboration. Consequently, in the joint lessons there was a presence of all subsets of the model in Figure 5.7, including sequences with a focus mainly on construction work, as well as with a focus mainly on mathematics. We share the concerns that Gellert and Jablonka (2009) address, about how mathematical reasoning may become neglected, or appear irrelevant, due to the presence of non-authentic reality in the mathematics classroom. However, in the case of this project, the vocational context was mainly authentic. We also agree with Gellert and Jablonka that sometimes it is preferable to stay within a mathematical context in the teaching. In this study, however, the aim was to connect mathematics to authentic workplace activities when relevant.

We see a danger with a form of teaching where what is counted as relevant knowledge (recognition rules, according to Bernstein 2000), and ways of expressing this knowledge (realisation rules) are not made explicit to the students. In such cases, students from homes with 'well"-educated parents will be likely to succeed more easily than students from homes with less well-educated parents. Most of the students in this study belonged to the latter group of students. Through our explicit work in making clear to ourselves (teachers, students, researcher) how relationships between mathematics and construction work may be characterised, the implicit was in fact made explicit. The model we developed was essential to this. The students became more and more engaged in the research project, and simultaneously in the teaching activities, both from a mathematics perspective and from the perspective of construction work. In other words, the project was both about keeping the two school subjects apart, *and* to integrate them in relevant ways. The project was also about giving students who are often excluded within the traditional system of schooling new possibilities to understand what counts as relevant knowledge, in school as well as in future workplace activities.

A limitation of this chapter is that we have not described the theoretical foundations of Bernstein (2000) in depth, but rather focused on aspects of concepts we adopted in terms of what was needed for this particular action research project. We refer to FitzSimons and Boistrup (2017; see also FitzSimons 2014a, 2014b) for further reading on theoretical perspectives on mathematics as recontextualised into workplace contexts.

The contribution to teaching and learning as well as to research is mainly through the model we developed, and also how we nuanced the understanding of the model through analysis of participants' answers to written questions. The number of participants was not high. Still, we contend that the model as such (as described also in FitzSimons and Boistrup 2017) may be viewed as a first idea of how interfaces between teaching content which characterise study programmes (such as construction work in vocational education) *and* mathematics may be understood. In future projects, the model may be used, and thereby modified according to each particular context. The model can be used to gain an understanding of how teachers who collaborate while crossing borders between teaching content can join in a collaborative teaching programme. Furthermore, the model shows how the quality of relationships between mathematics and vocational subjects does not primarily concern a specific location, but rather different kinds of teaching content.

Acknowledgements

We would like to express our gratitude to the city of Malmö, Sweden, which funded the project described in this chapter. Lisa's work is also funded by Stockholm University, especially in writing this chapter. We also would like to express our gratitude to Jonas Dahl, who read an earlier version in Swedish and who gave valuable comments, and to Gail FitzSimons who read the chapter close to its final version and provided us with important feedback on content and language matters.

Note

1   The findings of this chapter are also presented in Swedish in Bellander, Blaesild, and Boistrup (2017).

References

Atweh, B. (2005) "Understanding for changing and changing for understanding: Praxis between practice and theory through action research in mathematics education".

In Valero, P. and Zevenbergen, R. (eds), *Researching the socio-political dimensions of mathematics education: Issues of power in theory and methodology* Dordrecht: Kluwer Academic Publishers.

Bakker, A. and FitzSimons, G. (eds) (2014) "Characterising and developing vocational mathematics knowledge" [Special issue] *Educational Studies in Mathematics*, 86(2): 151–156.

Bellander, E., Blaesild, M. and Boistrup, L.B. (2017). "Matematik i yrkesprogram". *Forskning om Undervisning och Lärande 5*(2): 47–71..

Bernstein, B. (2000) *Pedagogy, symbolic control and identity: Theory, research, critique* (revised edn) Lanham, MD: Rowman & Littlefield .

Boistrup, L. B. and Gustafsson, L. (2014) "Construing mathematics-containing activities in adults' workplace competences: Analysis of institutional and multimodal aspects" *Adults Learning Mathematics: An International Journal*, 9: 7–23.

Boistrup, L. B. and Keogh, J. (2017) "The context of workplaces as part of mathematics education in vocational studies: Institutional norms and (lack of) authenticity". In Dooley, T. and Gueudet, G. (eds) *Proceedings of the Tenth Congress of the European Society for Research in Mathematics Education*, 1–5 February, Dublin: DCU Institute of Education and ERME.

Carr, W. and Kemmis, S. (2003) *Becoming critical: Education, knowledge and action research* London: Routledge.

Dahl, J. (2014) "The problem-solving citizen" Licentiate thesis, Malmö University.

Damlamian, A. and Sträßer, R. (2009) "ICMI Study 20: Educational interfaces between mathematics and industry" *ZDM Mathematics Education* 41: 525–533.

Dowling, P. (1998) *The sociology of mathematics education: Mathematical myths/pedagogic texts* London: Falmer Press.

FitzSimons, G. E. (2002) *What counts as mathematics?: Technologies of power in adult and vocational education* Dordrecht: Kluwer Academic Publishers.

FitzSimons, G. E. (2014a) "Commentary on vocational mathematics education: Where mathematics education confronts the realities of people's work" *Educational Studies in Mathematics*, 86: 291–305.

FitzSimons, G. E. (2014b) "Mathematics as vocational knowing: The importance of recontextualization" *Quaderni di Ricerca in didattica,* 24: 102–109.

FitzSimons, G. E. and Boistrup, L.B. (2017) "In the workplace mathematics does not announce itself: Towards overcoming the hiatus between mathematics education and work" *Educational Studies in Mathematics*, 95: 329–349.

Frejd, P. and Bergsten, C. (2016) "Mathematical modelling as a professional task" *Educational Studies in Mathematics*, 91: 11–35.

Gellert, U. and Jablonka, E. (2009) "I am not talking about reality". In Verschaffel, L., Greer, B., Van Dooren, W. and Mukhopadhyay, S. (eds), *Words and worlds: Modelling verbal descriptions of situations* Rotterdam: Sense Publishers.

Johansson, M. C. (2014) "Counting or caring: Examining a nursing aide's third eye using Bourdieu's concept of habitus" *Adults Learning Mathematics*, 9: 69–84.

Keogh, J. J., Maguire, T. M. and O'Donoghue, J. (2016) "Re-contextualising mathematics for the workplace" presented at 13th International Congress on Mathematical Education, July 24–31, Hamburg.

Lerman, S. (2000) "The social turn in mathematics education research". In Boaler, J. (ed), *Multiple perspectives on mathematics teaching and learning* Westport, CT: Ablex Publishing.

Lindberg, V. (2003) "Learning practices in vocational education" *Scandinavian Journal of Educational Research*, 47: 157–179.

Lindberg, L. and Grevholm, B. (2011) "Mathematics in vocational education: Revisiting a developmental research project, analysis of one development research project about the integration of mathematics in vocational subjects in upper secondary education in Sweden" *Adults Learning Mathematics*, 6: 41–68.

Muhrman, K. (2016) "Inget klöver utan matematik: En studie av matematik i yrkesutbildning och yrkesliv". PhD thesis, Linköping University.

Skolverket (2011) *Läroplan, examensmål och gymnasiegemensamma ämnen för gymnasieskola 2011*. Stockholm: Skolverket.

Wedege, T. (1999) "To know or not to know mathematics, that is a question of context" *Educational Studies in Mathematics*, 39 205–227.

Wheelahan, L. (2009) "The problem with CBT (and why constructivism makes things worse)" *Journal of Education and Work*, 22: 227–242.

Williams, J. S. and Wake, G. D. (2007) "Black boxes in workplace mathematics" *Educational Studies in Mathematics*, 64: 317–343

PART II

Mathematics education and everyday numeracies

Theoretical resources for analysis

Keiko Yasukawa

The four chapters in this Part illustrate the use of different resources that can be drawn upon to study interactions (or lack thereof) between mathematics education and people's everyday numeracies. The authors are concerned about the traditional classroom teaching and assessment regimes that give scant attention to what the learners can bring to the learning from their own lives, and how they are making sense of what is being taught to them.

A widely used type of mathematical task is the 'word problem' – mathematical problems that are dressed up in a 'realistic' context by a short narrative. In Chapter 6, Barwell adds to research on mathematical word problems by drawing on the work of Bakhtin. He shows that despite the authority conveyed by word problems in mathematics textbooks, and the normative practices that students are expected to deploy in order to solve the problems, students' own realities of life necessarily enter into dialogue as they read and try to interpret one of these problems. A word problem, Barwell shows, is a 'social text' that reflects tensions between the forces that steer students towards the normative practices of this class of academic problems, and the forces of students' life experiences that pull students away from the academic arenas. Understanding the tensions that emerge, Barwell argues, is critical to understanding how word problems create social stratification between those who are 'at home' with the normative practices and those who are not.

The question of exclusion from classroom discourses is picked up again in Chapter 7. Knijnik and Wanderer draw on Wittgenstein's idea of language games to contrast the language games of formal and abstract mathematics that are privileged in the curricula and classrooms, and the language games that are practised outside the school walls in people's lives. They argue that the lack of connection between the language games of the privileged academic mathematics and the language games of people's life-critical ethnomathematics needs to be challenged. They further argue that both types of language games require pedagogical attention, and call for educators and education researchers to seek ways of building a different kind of curricula.

The privileging of the kind of mathematics that these two chapters problematise is a phenomenon that travels across cultural and national boundaries. In Chapter 8, Ashwaikh and Yasukawa examine this phenomenon in the context of Occupied Palestine. They observe, through Ashwaikh's account of mathematics education in this region, how not only the kind of mathematics but also the pedagogical approach that reproduces the ideology surrounding this kind of mathematics are transported by the adopted textbooks. Students' lack of engagement seen elsewhere in the world is also reproduced. The authors consider whether increasing student-talk in the pedagogical design might be the key to challenging the silencing of student voice and agency that is currently being supported by and through the textbooks.

The cultural and historical influences of widely disseminated and adopted curricula and pedagogies on local practices is seen again in Chapter 9, in Shiohata's study of primary teacher training in Nepal. Shiohata reports on her work with teachers in a region in Nepal on using more learner-centred pedagogies and continuous assessment as ways that allow teachers to better understand how their students are learning. She reports on the difficulties of shifting teachers' practices from the long-established teacher-centred pedagogies and a view of summative assessment as the only form of valid assessment. However, Shiohata observes how shifts in pedagogies, where they do occur, are helping to reveal important information. Key to these changes are 'outside' agents like Shiohata who are working with the teachers, observing their classroom practices and working with the resources, however limited, afforded by their particular contexts.

Together, these chapters illustrate how problematising the interactions between mathematics education and everyday numeracies requires analysing and disturbing historically dominant educational practices. Linguistic, philosophical and pedagogical theories are some of the resources deployed by the authors for analysis and explorations of possible change.

6

WORD PROBLEMS AS SOCIAL TEXTS

Richard Barwell

An age-old problem

Word problems have long been a staple of mathematics classrooms around the world. Word problems involve a mathematics task being embedded within a written 'real world' context and are often justified as a way to encourage the application of mathematics to real situations. They have also been criticised for being too artificial (e.g., Verschaffel, De Corte and Lasure 1994; Verschaffel, Greer and De Corte 2000).

Word problems have existed since at least Babylonian times, more than 3500 years ago – Babylonian clay tablets include word problems designed to train scribes in particular kinds of calculation, such as fractional arithmetic in the context of weights and measures (Open University n.d.). Contemporary word problems are instantly recognisable, suggesting they have a clearly established, widespread generic form. The average school student in most education systems around the world will solve hundreds of such problems during their school career. They thus develop a clear sense of how word problems are organised. In some of my own research, I asked primary school students in the UK to work in pairs to write arithmetic word problems (see Barwell 2005 for more details). Here are some examples of what they came up with:

1   It was Christmas Eve and 20 people came to the Christmas party and there was a big cake. How many pieces do they get each?
2   There are 100 cars in the town car park. There are 4 floors. How many cars on each floor?
3   There are a hundred and eighty brains in the morgue. Four monsters came and eat 44 each. How much is left?
4   I have 150 cars in my business. Group these into 3 groups to see if you're clever.

These student-generated problems illustrate how children have a pretty clear idea of what constitutes a word problem. The first has the right elements (a situation, some numerical information, a question), but is ambiguous. All four problems, particularly the third one, suggest that the kinds of contexts students are interested in may be rather different from the humdrum affairs of apples, pencils and buses that seem to populate many textbook word problems. The fourth problem gives a good insight into how students may perceive word problems: not as an application of mathematical knowledge, but as a test of intelligence.

Indeed, research suggests that successfully solving word problems can be challenging. Educationalists have sometimes expressed perplexity at the frequency and nature of students' errors, particularly when these errors involve apparently nonsensical assumptions or solutions (e.g., Greer 1997). Moreover, there is evidence that students' backgrounds are a factor in how successfully they respond to such problems, implying that word problems are not so much a test of mathematical understanding, as a test of who the student is (e.g., Cooper and Dunne 2000). Students from working-class backgrounds, minority-cultural backgrounds or minority-language backgrounds may all perform, on average, less well on word problems than students from middle-class, cultural-majority backgrounds (Cooper and Dunne 2000).

So how do different students respond to word problems? How do word problems work in mathematics classrooms? How can a form of mathematics problem result in differential outcomes for students of different backgrounds? In this chapter, I argue that in order to answer these questions, it is necessary to examine word problems as social texts. That is, word problems must be understood not simply as neutral conveyances for mathematical tasks, but as socially constructed, deployed and interpreted texts. My thinking draws particularly on theoretical ideas from the work of Mikhail Bakhtin, and is illustrated with examples from elementary school mathematics classrooms in Canada. These various ideas are described in the sections which follow. First, however, I provide a summary of some key ideas identified in the research literature on word problems.

Word problem research

Research has been conducted on children's performance on word problems since the beginning of research in mathematics education in the 1970s. This early attention to word problems reflects the widespread experience of teachers, curriculum specialists and researchers that students often struggled to solve word problems. Of particular concern was the observation that students often gave 'nonsensical' answers to word problems (e.g., Greer 1997; Verschaffel, De Corte and Lasure 1994).

Much early research sought to determine if differences in the mathematical structure of a word problem influenced students' performance. In this approach, researchers developed typologies of word problems based on the mathematical transformations they embodied, and then compared students' performance on the different kinds of problem (see, for example, Carpenter, Hiebert and Moser 1981; Carpenter and Moser 1984; Vergnaud and Durand 1976). Addition problems, for example, can be divided into two types: join and combine problems. Join problems involve a quantity being added on to an initial quantity:

> Wally had 3 pennies. His father gave him 8 more pennies. How many pennies did Wally have altogether?
>
> *(Carpenter and Moser 1984, 180)*

Combine problems involve two initial quantities being combined together:

> Sara has 4 sugar donuts. She also has 9 plain donuts. How many donuts does Sara have altogether?
>
> *(Carpenter and Moser 1984, 180)*

Carpenter and Moser (1984) investigated whether the mathematical transformation required in a problem influenced children's solution strategies. They found no difference between strategies used by young children for join or combine problems, although these strategies changed as children progressed from first to third grade. These strategies involved either counting out first one addend and then continuing the count through the second addend, or, subsequently, counting on from the larger addend (Carpenter and Moser 1984, 181, 190–192). Differences were noticed, however, in children's strategies for different kinds of subtraction problem. Depending on the structure of the problem, children might count up from one number to another, count down from one number to another, or construct a set of objects representing the larger number and then remove the smaller number of objects to model the subtraction (see Carpenter and Moser 1984, 182, 192–193). This kind of research has continued for many years, with various studies looking at differences in problem structure, wording or contextual information as a way to explain differences in performance or solution strategy. Reusser and Stebler (1997) provided a good summary of the main findings of this body of research:

- students frequently solve problems without understanding them;
- students readily 'solve' unsolvable, even absurd, problems if presented in ordinary classroom contexts;
- students almost never ask themselves if a problem given to them is solvable or not;
- students frequently use superficial key word methods (or direct translation strategies) rather than thinking deeply about the implied real-world situation when solving stereotyped [sic] word problems;

> • students' factual problem-solving behavior is heavily influenced by contextual information;
> • variations in the 'presentational structure' of tasks (changes in wording) dramatically affect problem difficulty;
> • students who can easily deal with additive and subtractive problems within the classroom seldom use the formal arithmetic notations when asked to write down what happened in real-world situations dealing with candy, flowers or dice.
> *(Reusser and Stebler 1997, 310, extensive supporting references omitted)*

In response to this broad set of issues, a more recent strand of research has sought to develop models of the word problem-solving process in a bid to understand where in the problem-solving process students failed to use realistic considerations and design interventions to correct these failures. This work draws on a 'modelling' perspective: that is, the aim is for word problems to be treated as part of a more general class of realistic problem-solving. The 'standard' account of this problem-solving process involves several cyclically related phases:

> • understanding the situation described;
> • constructing a mathematical model that describes the essence of those elements and relations embedded in the situation that are relevant;
> • working through the mathematical model to identify what follows from it;
> • interpreting the outcome of the computational work to arrive at a solution to the practical situation that gave rise to the mathematical model;
> • evaluating the interpreted outcome in relation to the original situation;
> • and communicating the interpreted results.
> *(Verschaffel, Greer and de Corte 2000, xii)*

As Verschaffel, Greer and de Corte point out, this general account of the modelling process can be applied to word problems (136). Based on their extensive research, however, they propose a more complex version of the modelling process is necessary to account for the varied ways in which students may tackle word problems within the social context of a mathematics classroom. From this perspective, students' errors in word problem solving are not due to intellectual deficits, but due to the social context in which word problems are introduced and used (i.e. they are presented as repetitive exercises, are designed to test specific mathematical operations, and have a single correct answer). Verschaffel, Greer and de Corte's elaborated model embeds the modelling phases summarised above within a broader social context. This broader context includes:

- students' (access to) knowledge about the phenomenon under investigation, such as through personal experience, knowledge of peers, information from books and other sources;
- students' attention to the goals of the problem, such as with respect to the degree of precision required, the number of contextual factors that should be taken into account, etc.;
- resources available to students, such as mathematical techniques;
- the possibility of multiple models for a given situation and the resulting need to compare different models and make a selection;
- the technical requirements of the task, particularly with respect to the communication of students' work and their solution.

(based on Verschaffel, Greer and de Corte 2000, 168–171)

As a result of their research, Verschaffel, Greer and de Corte argue for a more realistic approach to the use of word problems in mathematics classrooms. In particular, they propose that instruction needs to allow for the broader contextual components of mathematical modelling enumerated above. For example, students need to be encouraged to consider a range of sources of information, consider different possible models and mathematical techniques, and make informed decisions about how to communicate their work.

Finally, a somewhat distinct line of research has examined the differential performance of students on word problems in relation to various background social factors, including home language, race and social class (e.g., Cooper and Dunne 2000; Cooper and Harries 2005; Nunes, Schliemann and Carraher 1993; Secada 1991). These studies have shown that students from some backgrounds, such as, in particular, lower socioeconomic groups, are more likely to give unrealistic responses than others, or conversely, apply realistic considerations inappropriately. Cooper and Harries (2005, 150), for example, interviewed 55 primary school children about a problem they had previously encountered in a test-like situation (see Figure 6.1):

Students are expected to complete a division calculation and then round the remainder up, since one extra trip is required to carry the left over people. This

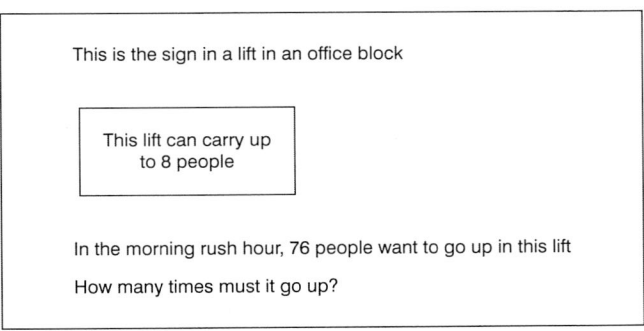

FIGURE 6.1   Division word problem (Cooper and Harries 2005, 150)

problem is a version of a classic situation in which a realistic consideration is required to interpret the remainder; a common error is to give the response 9.5 or even 9. On the other hand, students are *not* expected to be too realistic; they should not assume that the lift is not always full, for example. Cooper and Harries found that working-class participants were more likely to choose the wrong operation or to treat the remainder inappropriately. They also found, however, that suitable encouragement, including alternative versions of the problem, could prompt students during interview to pay more attention to realistic considerations.

Cooper and Harries's (2005) findings are illustrative of similar results that link students' backgrounds to, specifically, their use of realistic considerations in word problem solving in school. Nunes, Schliemann and Carraher (1993), for example, compared Brazilian street children's mathematical problem solving in school and during commercial transactions conducted during their work as street vendors. They found, among other things, that the children solved problems much more accurately in the context of the commercial transactions than when they responded to similar problems in a formal test. This finding suggests that the more meaningful context of the children's everyday work activities was a factor in the accuracy of their responses.

Research suggests, then, that performance on word problems is related, in part, to students' backgrounds. In other words, word problems stratify students into different groups according to these backgrounds. While most research in this area is based on students' performance on tests, some useful insights have arisen from studies in which students are interviewed to explain their interpretation of problems and their reasoning while solving them. This kind of approach links individual thinking to broader social patterns, but tends not to interrogate the social nature of word problems themselves. In this chapter, I draw on the concept of genre as a way to explore word problems as social texts.

Word problems as genre

As a starting point, a genre can be defined as a recognisable stable pattern of textual organisation. A menu, for example, is generally organised in a particular way, with sections for different parts of a meal, or different types of dish. This example immediately suggests a relationship between the way a generic text is organised and its function within a given social situation. Menus are organised so as facilitate for the customer the process of selecting and ordering food and drink. Genres can thus be better thought of not simply as textual forms, but as social in nature. The function of mathematical word problems seems, superficially at least fairly straightforward. They are designed to facilitate students' practice in the application of certain mathematical techniques in situations derived from real life. A more thorough analysis, however, troubles these assumptions.

Gerofsky (1996) conducted an exhaustive analysis of word problems as a genre, and included examples from ancient Babylon and Egypt, sixteenth-century England and 1970s USA, as well as (then) contemporary examples from the 1990s. According to Gerofsky's analysis, word problems have three components. The first component is what Gerofsky calls the *set-up*, in which a context or situation is described and participants are introduced. For example: "brains in the morgue" is the set-up for the third student-generated problem introduced at the beginning of the chapter. Gerofsky points out that the information in the set-up is often not necessary for obtaining a solution to the problem. The second component is the *information* needed to complete the required calculation. In the morgue problem, the information is that there are 180 brains and that four monsters eat 44 brains each. The students' examples illustrate that the set-up and the information can to some extent be intertwined. Finally, the third component is the *question*: "How much is left?"

One of the puzzles about the word problem genre is that these problems overtly draw on 'real life' situations as a context for mathematical calculation and problem solving, and yet they are quite clearly *not* realistic. This observation has led several authors to wonder what word problems are actually for:

> All this leads me to a question for which I have no answer as yet, the question of the purposes of word problems as a genre. They are currently used as exercises for practicing algorithms, but such practice could certainly be achieved without the use of (throwaway) stories. The claim that word problems are for practicing real-life problem solving skills is a weak one, considering that their stories are hypothetical, their referential value is nonexistent, and unlike real-life situational problems, no extraneous information may be introduced. Nonetheless, they have a long and continuous tradition in mathematics education, and that tradition does seem to matter.
>
> *(Gerofsky 1996, 41)*

If, as Gerofsky suggests, word problems are simply part of the tradition of school mathematics, we can still ask what this tradition serves. Lave (1992) suggests that word problems in fact model an important aspect of mathematical thinking: the abstraction of a few salient features from the messiness of real life.

> The very process of solving word problems takes its form directly from the theory of learning: abstracting out the numbers and operations from a situation, operating on them in abstracted form, drawing a conclusion or generalization about the results, then reinserting the results into the situation. And finally, learning what arithmetic word problems are about, when those everyday scenarios are in fact greatly stylized and distanced in other ways from the scenes and experiences they depict, is a sustained process during which children absorb a genre of puzzle-solving activity. Their experience with this genre teaches them to give different meaning and value to certain experiences in school and to other aspects of their

lives, including what they come to call 'real' math and the 'other' math of everyday activity.

(Lave 1992, 76)

Hence, if word problems are a form of 'traditional' mathematical activity, their purpose is, like all traditional activities, to transmit certain practices, ways of seeing and values. The work of Verschaffel, Greer and de Corte (2000), discussed earlier in this chapter, can be seen as an attempt to change both the function and the tradition of word problems, so that their use more closely resembles problem-solving in the 'real world'.

If word problems are a form of tradition, however, Bakhtin's (1986) ideas about genres are helpful. Bakhtin makes a distinction between primary or simple genres, and secondary or complex genres:

> Secondary (complex) speech genres – novels, dramas, all kinds of scientific research, major genres of commentary, and so forth – arise in more complex and comparatively highly developed and organized cultural communication (primarily written) that is artistic, scientific, sociopolitical, and so on. During the process of their formation, they absorb and digest various primary (simple) genres that have taken form in unmediated speech communion. These primary genres are altered and assume a special character when they enter into complex ones. They lose their immediate relation to actual reality and to the real utterances of others.
>
> *(Bakhtin 1986, 62)*

Word problems are a form of secondary genre, primarily written in the context of the specific quasi-scientific activity of mathematics instruction. The 'real-life' contexts that form the set-ups of word problems arise from primary speech genres. A nice example can be seen in Cooper and Harries's (2005) lift problem: the problem presents a 'sign in a lift' that says, 'This lift can carry up to 8 people'. The problem incorporates one genre (notices) into the secondary genre of the word problem. Moreover, Bakhtin's next point in the above quotation is significant: the notice in the lift is altered and 'assumes a special character' within the word problem. The notice is no longer a notice, but a piece of relevant information to be used to solve a mathematical problem. This change of character illustrates well Bakhtin's point that primary genres 'lose their immediate relation to actual reality'.

The ways in which primary genres are transformed within secondary genres is central to Bakhtin's view of (secondary) genres as ideological. Secondary genres impose particular ways of seeing or worldviews through the way in which they combine and reinterpret primary genres. Hence, the incorporation and transformation of the notice in the lift within the word problem reflects the particular worldview, or ideology, of word problems. This ideology concerns not simply a way of looking at the outside world, of distinguishing between what Lave called '"real" math and the "other" math of everyday activity, (Lave 1992, 76); it also concerns the position of the students themselves in relation to the world depicted in the word problem.

Genres should not, however, be seen as monolithic conveyances for static worldviews (Briggs and Bauman 1992). Any text is part of a continual dynamic process of reproduction, interaction and adaptation. Gerofsky's characterisation of word problems as a genre is not an exercise in uncovering the eternal rules of this particular form of text; rather, she has identified certain patterns arising through a centuries long process of continual reinvention of word problems. Babylonian word problems are not exactly like the problems found in contemporary mathematics textbooks; each reflects the requirements of their time and place. And yet there are some identifiably similar features in word problems over time.

In the rest of this chapter, I expand on this social account of word problems, drawing more extensively on interrelated ideas from Bakhtin's work. In particular, I discuss: word problems as *dialogic*; word problems and *heteroglossia*; and word problems and *stratification*.

Word problems as dialogue

The notion of dialogue is a complex one that includes several related aspects. Dialogue is about a kind of relation, such as, fundamentally, the relation between utterances or texts. In any exchange of utterances, what is said anticipates an *addressee*; another way to say this is that what is said is shaped to suit the intended audience (Bakhtin 1981). Word problems certainly reflect this aspect of dialogue. They are clearly designed for a particular audience – mathematics students – and, as I have discussed, the genre reflects (and constructs) this audience. For example, the question component of a word problem is directly addressed to the student receiving the problem; it is, indeed, an instruction, as much as a question, reflecting the particular pedagogic relations of a mathematics classroom.

Any utterance also expects and generates a response. Word problems anticipate a particular kind of response: the expected answer. But any utterance prompts a wide range of responses. The student-generated word problems quoted at the start of this chapter give some insights into students' dialogic interaction with the word problem genre. I have already pointed out that the students seem to be quite familiar with the main features of the genre. A closer look shows that the students have also appropriated aspects of the ideology of word problems. Most explicitly, the injunction to 'see if you're clever' displays a keen awareness of the de facto evaluative function of word problems. There are other examples, however, such as the unstated assumption in the car park question that cars will be distributed evenly between the four floors. Meanwhile, the morgue question can be heard as an ironic comment on the blander contexts of word problems found in textbooks.

This appropriation of the generic language of word problems into their own utterances reflects another aspect of dialogue: the internal interaction between different voices. In the student-generated problems, we hear the voices of both students and traditional mathematics textbook problems. It is the interaction between these voices that generates the humorous effect of some of the problems. Of course, our own voices as readers interact with these other voices to also contribute to this humorous effect.

These various aspects of dialogic relations can be seen in the following example, in which two third-grade students tackle a word problem given on a worksheet in their French immersion class. The problem is written in French, although much of the students' interaction is in English.[1] The problem is as follows:

> Joseph boit un demi litre de lait par jour. Combien de jours est-ce qu'il faut pour finir quatre litres de lait? [*Joseph drinks a half litre of milk each day. How many days are needed to finish four litres of milk?*]

Kyle:	okay what the heck is this? (*reading*) Joseph boit un demi-litre de lait par jour (.) combien de jours (.) [*Joseph drinks a half-litre of milk each day (.) how many days (.)*]
Sara:	(un deux trois) [(*one two three*)]
Kyle:	aaahhh (.) so four litres is like up to here four litres is two sizes of this (.) if he drinks that (.)
Sara:	(*inaudible*)
Kyle:	on my calcumalations (.)
Sara:	but we're already at two
Kyle:	oh I know what he (.) I know what this is (.) it's um ah ah ah um (.) grrr (.) I know it's something around (.) to drink (.) 'kay Joseph (.) oh demi-litre (.) two c'est deux um ah litres (*writing*) deux litres (.) just do two and a capital L
Sara:	combien de jours [*how many days*]
Kyle:	no no (.) if it's a (.) if he does (.) how, if he does a deux a half of a milk carton a day (.) if he does one (.) like you know how you have those you guys have those you know how we have those mini
Sara:	yeah
Kyle:	milk cartons? that's a half a demi-litre (.) if you drink four of those how much is it?

Kyle starts to read the problem aloud, an act that doubtless occurs in mathematics classrooms around the world. Reading out a printed word problem necessarily and automatically involves two voices in dialogue: the voice of the word problem text and Kyle's voice. The textbook is in some sense an official or authoritative voice (Bakhtin 1981) – it is not meant to enter into dialogue with anything, but to be followed. Ultimately, however, all texts become dialogic the moment they are read. Kyle inflects his reading of the problem with his ideas, interpretations and accents. He prefaces his reading with the arch question 'what the heck is this?', thus already framing his reading with a degree of levity. His French has a strong anglophone accent.

The two students start to vocalise their interpretations, partly in the form of calculations conducted out loud, but then Kyle explicitly offers an interpretation of the problem: 'I know what this is (.) it's um ah ah ah um (.) grrr (.) I know it's something around (.) to drink.' His identification of 'to drink' as relevant information suggests that at this point he is clarifying the meaning of some of the information in the problem, such as, that 'boit' means 'drink'. Similarly, he offers an interpretation of another aspect of the problem 'we have those mini milk cartons? that's a half a demi-litre', in which he makes sense of 'demi-litre', relating it to a particular size of milk carton. So far, Kyle has constructed a version of the set-up; he then interprets the question: 'if you drink four of those how much is it?', which fits with his proposed solution of two litres. We can thus reconstruct Kyle's version of the problem.

> Joseph drinks a half litre carton of milk each day. If you drink four of those how much is it?

This example gives a good sense of the meaning of a word problem emerging through the interaction between the text, Kyle's readings of the text, and his experience of, in this case, milk cartons. Half a litre of milk, for example, becomes a half-litre *carton* of milk. Multiple voices are in dialogue and Kyle's version of the problem emerges from this dialogue.

The generic nature of word problems contributes to Kyle's interaction with the problem. Even though Kyle ends up solving a different problem to the one stated, he nonetheless works at interpreting the set-up, and then at interpreting the question. He clearly understands the purpose of the word problem in the context of the class's mathematics topic on capacity and displays appropriate behaviour in response to the genre, such as reading the problem aloud, interpreting the information appropriately (he doesn't ask who Joseph is, for example), seeking a suitable calculation to perform and producing a generically appropriate solution.

Word problems and heteroglossia

The concept of heteroglossia, developed particularly in Bakhtin's (1981) analysis of the language of novels, is about the diversity inherent within language at every level, from the basic units of pronunciation to the nature of different natural languages. Despite the idea that different aspects of language are fixed and governed by strict rules (for example, of grammar, of pronunciation, of spelling), this unitary view does not reflect the creative, messy, inconsistent nature of language-in-use in social context. Heteroglossia is apparent in the word problem genre, as in any genre. While Gerofsky's (1996) work in particular has helped to clarify common features of word problems, including the main structure, as well as common grammatical and referential forms, it does not uncover 'rules' so much as tendencies. Word problems themselves are enormously varied, while still being recognisable as word problems. A look at mathematics textbooks around the world suggests that they vary from country to country, from topic to topic, from one language to another, from one teacher to another, and so on. This sense that word problems are varied but still recognisable is explained by Bakhtin (1981) in terms of the metaphor of centripetal and centrifugal forces. The centripetal forces push language in a normative direction; so, if you're writing a word problem, it should fit the genre as closely as possible. The centrifugal forces, however, push language in the direction of diversity; every word problem is unique and reflects a unique sociocultural context, such as, for example, the language, age of students, topic, curriculum requirements, teacher's sense of humour, or current topics of interest. Bakhtin's key insight is precisely that these two forces are always in tension in language, so that any utterance, such as a word problem, reflects both. So the morgue problem, for example, reflects perfectly both the expected requirements of a primary school arithmetic word problem, and the individual imagination and relationships between the author and his peers that led to the production of this particular word problem.

Heteroglossia also shapes students' interactions with word problems, particularly within their pedagogic context. The genre not only structures the text of the problem, but the organisation of interaction relating to the text. Minimally, students need to read the problem, interpret the information, perform a calculation and write down a solution. These requirements amount to a centripetal force that shapes how students respond. In the example featuring Kyle, for instance, he does indeed follow these various steps. Heteroglossia, though, is apparent throughout the episode. Kyle's utterances feature novel word forms, non-standard interpretations, accented pronunciation and the use of two different languages.

In the next example, observed in a Grade 6 class of Cree students, their (white) teacher talks through a problem designed to practise estimating. The problem shows pictures of four items with price tags: a camera, a laptop, a video-camera and something that might be a television. The problem then states:

> Suppose you have won a $5000 shopping spree at an electronics store.
> a   Estimate to find out if you could choose 1 of each item.
> b   If the total is too great, which item would you put back?
> *(From Maths Makes Sense, Grade 6, p. 216,*
> *problem slightly simplified, part c omitted)*

Teacher A: number three if you look down to number three in your textbook
(.) number three says and you are only doing A and B (.) number

	three says suppose (.) which means pretend (.) pretend you won a five thousand dollar shopping spree (.) pretend you have won five thousand dollars to spend on a shopping spree at an electronic store (.) so that means like Best Buy or Future Shop those types of stores (.) number A or A says estimate to find out if you can choose one of each item in question number two (.) so they want you to look at the prices in question number two and say okay could you buy one of those each (.) one camera one laptop one video camcorder and one TV and not go over five thousand dollars
Student:	nope
Teacher A:	say yes or no okay (.) and then B says (.) if the total is too great which item would you put back and why (.) so you're going to have to check and add all those up and check and see how much they would add to (.) if it's over five thousand if it's over what you have won (.) if it's more than what you have won (.) which item would you put back and why okay

This example illustrates both a multiplicity of voices, as well as a multiplicity of discourses in relation to the estimation word problem, both of which are aspects of heteroglossia (Busch 2014). The teacher, like Kyle, reads through the problem, but does so in a way that quite strongly intermingles her own voice with that of the textbook. For example, rather than simply reading 'suppose', she says 'suppose (.) which means pretend', thus introducing a synonym for suppose, which she then uses consistently. Similarly, she interprets 'electronic store' for the students 'so that means like Best Buy or Future Shop' (Best Buy and Future Shop are both well-known electronics stores in Canada). And she unpacks question (a) with a kind of thinking aloud: 'so they want you to look at the prices [...] and say okay could you buy one of those each [...] and not go over five thousand dollars'. This last unpacking explicitly refers to the author of the problem ('they') and the thinking aloud includes reworking the problem in language of a more everyday, oral tone ('okay'). This interweaving of voices and ways of talking, incorporating different authors (they, you), perspectives (pretend, what they want) and discourses (estimate, okay) is a good example of heteroglossia arising in response to a word problem. Moreover, the teacher's reading is clearly designed for her audience (struggling Grade 6 students) who she constructs, for example, as knowing about electronics, but not necessarily knowing what a shopping spree is. The requirements of the task are also apparent, such as in the teacher's 'they want you to', or in her second turn 'you're going to have to check', underlining the authoritative nature of the genre, even while softening the language. Indeed, the teacher's careful interpretation of the problem assumes that spending $5000 on electronics is not necessarily part of their everyday experience. Her interpretation perhaps seeks to make the problem more 'realistic', therefore: she refers to actual stores and to the specific items listed in the problem, even though, generically, these details are not important.

The students get to work on their page of problems, and, as each student reaches the shopping spree problem, they ask the teacher for help. Much of these interactions are not, in fact, about the realistic details of the problem, but about what to write, and where to write it on their answer sheet. Here's one example:

Teacher A: number three (.) so it says estimate to find out if you can choose one of each items [...] so (.) make a guess (.) can you do you think all these add up to a thousand or five thousand (.) or more (.) or less (3.0)

Student: I think less

Teacher A: less? okay (.) so (.) estimate if you can choose one of these items [...] do you think you could choose one of each item without going over? or do you think you'd go over (2.0)

Student: I won't go over

Teacher A: you won't go over? okay so then just write out I won't go over (.) so I won't go over five thousand dollars

The teacher and the student jointly produce a response to the problem. The teacher poses the key question directly to the student, transforming 'Estimate to find out if you could choose one of each item' into 'do you think all these add up to a thousand or five thousand'. In response to this modification in both register (from more formal mathematical to less formal everyday) and modality (from 'estimate' to 'do you think'), the student is able to indicate a position: 'less'. The teacher then guides him further with what to write, thus supporting him to provide a generically suitable response. One of the challenges here is that, ideally, students should learn not only the appropriate mathematical treatment to apply – they should also learn the generically appropriate way to formulate their response. It may be in relation to these double tasks that social differences may emerge, as discussed in the next section.

Word problems and stratification

The final aspect of Bakhtin's theory I want to introduce is his notion of stratification. In general, stratification arises as a result of centripetal and centrifugal forces. The idea that there are correct ways to use language is connected with social structure, since these ideas are normative (even if they are sometime presented as absolute). That is, while there are always many ways to say something, some ways are considered closer to the norm than others, and are therefore, in most cases, considered more socially acceptable, or even more 'correct'. In mathematics classrooms, for example, mathematical explanations that draw on recognised mathematical discourse are considered preferable to explanations that use everyday language (Barwell 2016).

Stratification arises specifically in relation to genres. Indeed, for Bakhtin, genres are a key means through which the stratification of language occurs:

> Literary language – both spoken and written – although it is unitary not only in its shared, abstract, linguistic markers but also in its forms for conceptualizing these abstract markers, is itself stratified and heteroglot in its aspect as an expressive system, that is, in the forms that carry its meanings.
>
> *(Bakhtin 1981, 288)*

Genres, then, while having idealised forms, are diverse in their expression, with this diversity of expression being stratified. Moreover, genres are deployed in a variety of different contexts, including different social or professional domains, just as a word problem is from the world of mathematics classrooms, but may also arise in a research article, or a novel, and so on. Briggs and Bauman develop this link between genres and the domains in which they are used:

> Genres also bear social, ideological, and political-economic connections; genres may thus be associated with distinct groups as defined by gender, age, social class, occupation, and the like. Invoking a genre thus creates indexical connections that extend far beyond the present setting of production or reception, thereby linking a particular act to other times, places, and persons. [...] Genre thus pertains crucially to negotiations of identity and power – by invoking a particular genre, producers of discourse assert (tacitly or explicitly) that they possess the authority needed to decontextualize discourse that bears these historical and social connections and to recontextualize it in the current discursive setting.
>
> *(Briggs and Bauman 1992, 147–148)*

Thus, proficient use of a genre, whether in producing appropriate examples, or responding appropriately to generic texts, is a social marker, indicating a degree of mastery not just of the content, but of the form, of the underlying worldview, and of the related social situation. It is this last point that may explain how word problems differentiate students according to their backgrounds.

In the shopping spree problem, for example, the teacher clearly (and not surprisingly) demonstrates a clear command of the genre. She is able to read, interpret, unpack and reformulate the text of the problem; she models possible thinking strategies, guides students towards mathematically appropriate responses and supports the students to produce generically suitable answers. Her students display much less sureness, whether in reading, offering ideas or developing full written responses. A narrow cognitive reading might conclude that the students have not acquired the necessary knowledge or do not have the ability to tackle such problems (a conclusion I reject).

Instead, it seems to me that word problems, to use a Bakhtinian term, are a form of *alien word* for the students: word problems represent another's worldview, which thereby alienates these students. This otherness may be apparent in the set-up of the problem. It is unlikely that elementary school students have much experience of spending $5000, although they all seem to be familiar with the electronics items. But the set-up of a problem is perhaps not the most significant issue. The worldview of word problems involves a view of both the real world, and of how to read the real world in order to work out what is being asked – the decontextualisation and recontextualisation referred to by Briggs and Bauman (1992).

This difference of worldview is apparent in the next extract, which comes later in the class working on the shopping spree problem. The teacher is working with two students on the second part of the question and they are discussing which item to remove in order to keep their spending below $5000:

Teacher A: so if you were to decide okay I'm gonna take off one thing what thing would you take off?
Student 1: the video camera
Teacher A: you would take the video camera?
Student 1: yes
Teacher A: why the video camera? (3.0)
Student 2: (*inaudible*)
Teacher A: no (.) because you don't want a video camera (.) which one would you take off?
Student 2: the camera
Teacher A: you would take off the camera why the camera? (2.0) why did you choose the camera why not the video camera? (2.0) why? (.) tell me why (.) I know you know the answer (.) why did you choose the camera to take off? (.) I'm going to write down camera for you? (.) but I need you to tell me why you chose the camera
Student 1: because (it takes video)
Teacher A: well maybe hey maybe it does you know maybe this is again an HD camera and you can take a video with it (.) maybe it's the same thing why would you take the camera over the video camera (.) I agree with [Student 2] but why? (.) why? (.) look at the price of the camera compared to the video camera
Student 1: that one's less this one's more
Teacher A: this one's less this one's more right so why would you want to take the camera over the video camera off your bill oh forget it Future Shop I don't want the camera I will take the laptop video camera and TV but I can't have the camera because I am over five thousand (.) why would you want to take the camera (3.0)
Student 1: it's less
Teacher A: it's less right (.) it costs less

This whole discussion can be characterised in terms of a difference of worldview. The two students are responding to the word problem realistically, in the sense that they propose to remove the item that appeals to them the least. The teacher, however, is looking for a mathematical justification. She indicates that a particular kind of response is required when she says: 'I need you to tell me why you chose the camera'. The 'need' arises from the requirements of the word problem (the genre) rather than for any actual personal need on the part of the teacher. In this pedagogical context, she is stating a requirement within the norms of school that students need to give reasons for the answers to mathematics problems. 'Why', in this case, is open to a few different possible responses. The students reply with realistic ones but their responses do not index the appropriate mathematical discourse. The teacher draws attention to a mathematical difference that can be made between the camera and the video camera: 'look at the price of the camera compared to the video camera', guiding the students to notice what for her is a salient distinction. The students are thus prompted to comment that one item is less than the other. The discussion continues, with the teacher taking the students through a calculation of the total when each item is separately removed, in order to show them that removing the camera leaves the total closer to $5000.

Stratification is occurring in this discussion, most noticeably in the difference of approach I have described. The students suggest non-mathematical reasons for their choices, but are thus not decontextualising the problem appropriately. That is, they are not reading the problem in a way that screens out personal preferences. During a year spent visiting this particular class (see, for example, Barwell 2014), I observed several similar situations. The students in the class were all Cree, one of Canada's indigenous peoples and, as such, historically oppressed and currently marginalised. These students attended an urban public-sector school, although they spent much of their childhood in remote communities where the Cree language is widely spoken. It is not possible for me to claim a direct link between their relatively marginalised status and their responses to problems like the shopping spree problem. Nevertheless, it is clear that the word problem genre belongs to a particular cultural tradition imposed in Canada through colonisation and the curriculum. In an encounter between the unitary nature of the genre and the heteroglossia of students' interpretations and responses, stratification occurs.

Discussion

My primary aim in this chapter was to argue that word problems need to be understood as social texts. Early research on students' performance on word problems generally treated them as mathematical texts or, at best, linguistic texts. This approach led to some issues that were difficult to explain, including, most notably, why it is that students so often give 'unrealistic' responses to word problems, and why it is that students' responses might be related to their socio-economic backgrounds.

Word problems can be seen as a kind of tradition (Gerofsky 1996) and, as such, as being a means through which a particular way of giving meaning to the world is passed on from one generation to another (Lave 1992). In the light of the Bakhtinian concepts I discussed later in the chapter, Lave's view makes a lot of sense. When she suggests that word problems 'give different meaning and value to certain experiences in school and to other aspects of their lives', she is describing the ideological dimension of word problems that, for Bakhtin, go with any (secondary) genre. Nevertheless, this view does not explain why students from some backgrounds are more successful in solving word problems (in the expected way) than others.

I have drawn particularly on Bakhtin's theory of language and, in particular, his ideas about genre, to highlight three different aspects of word problems as social texts: their dialogic nature, the way they reflect heteroglossia, and the way they reflect the stratification of all language. This exploration leads me to draw the following conclusions.

First, students' supposedly 'nonsensical' solutions to word problems are only really nonsensical because they are removed from their dialogic context. Much of my own research has, indeed, shown that students are capable of making sense of word problems drawing on a number of different resources, such as accounts of their own experiences, and their knowledge of mathematics, of languages and of the word problem genre itself (e.g., Barwell 2005). But this sense-making is a dialogic, interactive process, while a brief written response to a word problem is just one utterance, isolated from the preceding chain of utterances that led up to it.

Second, despite the best efforts of textbook authors and assessment item composers, multiple interpretations of any given word problem are always available. It is not therefore possible to write a problem that is not open to some degree of ambiguity. Moreover, these multiple interpretations are not inherent within the text of the problem; they derive from the *interaction* between the student and the problem. When Kyle worked on the milk problem, he thought of the small half-litre cartons of milk common in school canteens in Canada. Not all students will get milk in the canteen or have noticed the cartons, so their reading of the problem will be different.

Third, the inevitable variation in students' interpretations and responses results in stratification: some interpretations and some kinds of responses are considered more acceptable than others, both in terms of their mathematical content, but also the language used to present and support them. Unfortunately, these variations are likely to map onto socioeconomic differences, with students from backgrounds less exposed to certain kinds of decontextualising and recontextualising practices (the tradition) more likely to offer equally meaningful, but centripetally less appropriate responses.

The implications of these points for teaching and learning mathematics cannot be simplistic. If word problems are social texts, teachers need to engage with them as such. They need to find ways to allow students' interpretations to be included and support students to recognise differences in interpretations and to understand the how and why of what kind of responses are preferred within mathematical discourse. That is, rather than treating word problems as a mysterious tradition, full of secrets that only a lucky few will divine, they need to be seen as dialogic texts, to be discussed, deconstructed and adapted.

Acknowledgements

The data relating to the milk problem and the shopping spree problem was collected thanks to funding by the Social Science and Humanities Research Council of Canada, grant 410-2008-0544. Thanks also to Carrie Learned and Maya Shrestra for their assistance with data collection and transcription.

Note

1 Transcript conventions:
   Participants' words are represented in standard unpunctuated English and French orthography. Non-standard words or pronunciations are approximated also using standard orthographic conventions. Parentheses indicate where transcription is uncertain. Italic type within parentheses adds contextual information. Italic type within brackets provides my translation of participants' French into English. Short pauses are shown with (.) For longer pauses, length in seconds is provided e.g. (2.0).

References

Bakhtin, M. M. (1981) *The dialogic imagination: Four essays* Austin, TX: University of Texas Press.
Bakhtin, M. M. (1986) *Speech genres and other late essays* Austin, TX: University of Texas Press.
Barwell, R. (2005) "Working on arithmetic word problems when English is an additional language" *British Educational Research Journal*, 31(3): 329–348.
Barwell, R. (2014) "Centripetal and centrifugal language forces in one elementary school second language mathematics classroom" *ZDM Mathematics Education*, 46(6): 911–922.
Barwell, R. (2016) "Formal and informal mathematical discourses: Bakhtin and Vygotsky, dialogue and dialectic" *Educational Studies in Mathematics*, 92(3): 331–345.
Briggs, C. L. and Bauman, R. (1992) "Genre, intertextuality, and social power" *Journal of Linguistic Anthropology*, 2(2): 131–172.
Busch, B. (2014) "Building on heterglossia and heterogeneity: the experience of a multilingual classroom". In Blackledge, A. and Creese, A. (eds), *Heteroglossia as practice and pedagogy*, Dordrecht: Springer.
Carpenter, T. P. and Moser, J. M. (1984) "The acquisition of addition and subtraction concepts in grades one through three" *Journal for Research in Mathematics Education*, 15(3): 179–202.
Carpenter, T. P., Hiebert, J. and Moser, J. M. (1981) "Problem structure and first-grade children's initial solution processes for simple addition and subtraction problems" *Journal for Research in Mathematics Education*, 12(1): 27–39.

Cooper, B. and Dunne, M. (2000) *Assessing children's mathematical knowledge: Social class, sex and problem-solving* Buckingham: Open University Press.

Cooper, B. and Harries, T. (2005) "Making sense of realistic word problems: portraying working class 'failure' on a division with remainder problem" *International Journal of Research & Method in Education*, 28(2): 147–169.

Gerofsky, S. (1996) "A linguistic and narrative view of word problems in mathematics education" *For the Learning of Mathematics*, 16(2): 36–45.

Greer, B. (1997) "Modelling reality in mathematics classrooms: The case of word problems" *Learning and Instruction*, 7(4):.293–307.

Lave, J. (1992) "Word problems: a microcosm of theories of learning". In Light, P. and Butterworth, G. (eds), *Context and cognition: Ways of learning and knowing* Hemel Hempstead: Harvester Wheatsheaf.

Nunes, T., Schliemann, A. D. and Carraher, D. W. (1993) *Street mathematics and school mathematics* Cambridge: Cambridge University Press.

Open University (n. d.) "Babylonian mathematics" http://www.open.edu/openlearn/science-maths-technology/mathematics-and-statistics/mathematics/babylonian-mathematics/content-section-1.1 (accessed 5 February 2017).

Reusser, K. and Stebler, R. (1997) "Every word problem has a solution: The social rationality of mathematical modeling in schools" *Learning and Instruction*, 7(4): 309–327.

Secada, W. (1991) "Degree of bilingualism and arithmetic problem solving in Hispanic first graders" *Elementary School Journal*, 92(2): 213–231.

Vergnaud, G. and Durand, C. (1976) "Structures additives et complexité psychogénétique" *La Revue Française de Pédagogie*, 36: 28–43.

Verschaffel, L., De Corte, E. and Lasure, S. (1994) "Realistic considerations in mathematical modeling of school arithmetic word problems" *Learning and Instruction*, 4(4): 273–294.

Verschaffel, L., Greer, B. and de Corte, E. (2000) *Making sense of word problems* Lisse: Swets and Zeitlinger

7

BROADENING SCHOOL MATHEMATICS CURRICULUM

The complexity of teaching mathematical language games of different forms of life

Gelsa Knijnik and Fernanda Wanderer

Introduction

This chapter problematises the mainstream Western school mathematics curricula. Based on Wittengenstein's ideas, Knijnik (2012) argued that the mathematical practices generated by specific cultural groups can be understood as a set of language games associated with different forms of life, aggregating specific rationality criteria. The ideas presented by Wittgenstein are productive for problematising the understanding of a universal and foundationalist reason which sustains modern thinking, in which mathematics holds a privileged place (Walkerdine 1995). This productivity is one of the key points of interest for the ethnomathematics perspective. Using these ideas, we can say that the mathematical language games taught in mainstream Western school mathematics curricula follow the abstract and formal rules that constitute a specific grammar, associated with modern European rationality. As discussed elsewhere (Knijnik and Wanderer 2015), this grammar is marked by transcendence; therefore, it is understandable that there are great similarities among these school curricula. One of these similarities consists in their goal to leading the students to be able to solve everyday situations involving, for example, space, time, and quantification processes.

However, as we discuss here, everyday situations are necessarily solved taking into account their contingency. The mathematical language games practised by people to solve contingent situations are not precisely those taught at Western schools. Some of the examples given in this chapter show that they have some similarities. In Wittgenstein's perspective, we can say that there are family resemblances among them, but there are differences between school forms of life and outside forms of life.

Based on this argument, we show the complexity of teaching mathematical language games of diverse forms of life. Even so, we claim that it is necessary to broaden school mathematics curricula if we want the students effectively to learn to deal with their life necessities.

Setting the scene

> One puts the tractor on the land. If one works with it for three hours, you get one hectare.
>
> *(Knijnik 2000, 35)*

> I live very close to my work. It takes 10 minutes by bus to get there.

The first sentence of the above epigraph was an utterance made by a peasant during fieldwork performed at a Landless Movement (in Portuguese, Movimento Sem Terra) settlement in the south of Brazil years ago. His most important work was cultivating melons. As in other cultures, in order to define the quantity of seeds and fertilisers needed to plant them, it is necessary to demarcate the plot to be cultivated. In order to demarcate one hectare, a tractor was placed on the land. 'If one works with it for three hours, you get one hectare'. Here we are learning about a practice of using a unit of time to measure the land surface.

When the peasant first explained his calculation, we were surprised. It was only when we transcribed our conversation with him that we began to think about what had led to it. We perceived that in his practice, time and space were mixed: three hours was one hectare and one hectare was three hours. It is the tractor – or rather the costs involved in using it – that establishes a close relationship between time and space. From this view point, 'a few metres more, a few metres less, it makes no difference', as the peasant clearly said. Therefore, the hour of tractor use could possibly be a more relevant item of information than precision regarding an area of planted land.

This episode can be analysed in the light of the later writings of Wittgenstein. Those ideas express a concept of language no longer with the marks of universality, perfection and order, as though they existed before human actions. Wittgenstein's work challenges the existence of a universal language and the notion of total rationality, of an ontological foundation for language. Wittgenstein problematises the notion of a universal language and points to the constitution of various criteria of rationality. He assumes that language has a contingent and particular character, and that it acquires meaning through its various *uses*. 'The meaning of a word is its use in the language' (Wittgenstein 1958, 20).

The philosopher considers the existence of many languages which acquire meanings according to their use. He conceives language games as social practices

– to describe objects and analyse them, to tell stories, to 'do math in one's head' and to calculate the area of surfaces. To establish the meaning of the words, gestures and, one might say, the language games and their rationality, criteria are produced in the context of a given form of life. As Glock explains (1996, 124), the notion of 'form of life' expresses the 'intertwining of culture, world-view, and language'. Thus, the possibility of essences or fixed guarantees for a language is questioned, leading us to problematise the existence of a single mathematical language with fixed meanings.

As discussed elsewhere (Knijnik 2012), based on the later work of Wittgenstein and some of his interpreters (e.g., Glock 1996), it can be argued that different forms of life produce different language games, each marked by its specific grammar. Such grammar, as a set of rules, constitutes a specific logic. It entails more than a single language: there are different language games. Thus we can ask: Is there some kind of relationship between the different language games? If the answer is positive, how do they relate? The response to these questions was provided by Wittgenstein's later work through the notion of family resemblances. The philosopher considers (as shown in Aphorisms 66 and 67 of *Philosophical Investigations*) that language games form 'a complicated network of similarities overlapping and criss-crossing: sometimes overall similarities, sometimes similarities of detail' (Wittgenstein 1958, 32). The relationship between the language games and the family resemblances engenders the rationality criteria (ibid).

The episode with the peasant mentioned above refers to the practice of a language game associated with the peasant's working form of life. This game follows the rule: to determine the surface of one hectare by three hours of tractor operation. Through the notion of family resemblances, it is possible to characterise this language game as mathematical. This is because it bears resemblances to the language games transmitted by the mathematics in which we were schooled.

The mathematics language game practised by the peasant emerged as one among many examples of peasant forms of life in the south of the country. It was a form of life marked by material precariousness, where the small farmer's work rarely manages to compete with the large landowners'. Thus, the cost of production, particularly the amount to be paid for the leasing of the machinery, is a central point. These are the conditions that give meaning to the rule that associates time-space, which characterises this language game. In this context, the precision of the area to be cultivated is no longer relevant. It appears to be an opportune moment to return to the lessons given by Foucault (1972), particularly to the discussion he undertook about the notion of discipline as an internal procedure of discourse control. The disciplines constitute a sort of anonymous system, freely available to whoever wishes, or whoever is able to make use of them, without there being any question of their meaning or their validity being derived from whoever happened to invent them. For a discipline to exist, there must be the possibility of formulating fresh propositions – and of doing so *ad infinitum* (Foucault 1972, 222–223).

But the as-yet unformulated propositions still to be generated cannot be just any 'In every discipline, there are objects, methods, true propositions, rules, definitions, techniques and instruments to be used by its possible participants' (Díaz 2005, 80); propositions that are not aligned with these are considered spurious and should therefore be excluded from the discipline, 'but it repulses a whole teratology of learning' (Foucault 1972, 222). The philosopher explains this point:

> A discipline is not the sum total of all the truths that may be uttered concerning something; it is not even the total of all that may be accepted, by virtue of some principle of coherence and systematisation, concerning some given fact or proposition.
>
> *(Foucault 1972, 223)*

If, for Foucault (1972, 223), 'Medicine does not consist of all that may be truly said about disease', and the same goes for botany as regards plants, then we might think of extending this position to mathematics education and, paraphrasing the philosopher, say 'that [mathematics education] cannot be defined by the sum of all the truths that concern the language games involving quantification' (such as for instance, calculating the area of surfaces). For example, we would say that school mathematics does not bring together all language games to calculate areas, 'repelling outside-its-margins' games such as calculating the surface of a hectare using three hours on a tractor. This language game produces an approximate result; it is not exact.

However, as Wittgenstein points out in Aphorism 88 of *Investigations*, inexact does not mean useless. The philosopher writes:

> 'Inexact' is really a reproach, and 'exact' is praise. And that is to say that what is inexact attains its goal less perfectly than what is more exact. Thus the point here is what we call 'the goal'. Am I inexact when I do not give our distance from the sun to the nearest foot, or tell a joiner the width of a table to the nearest thousandth of an inch? No single ideal of exactness has been laid down; we do not know what we should be supposed to imagine under this head – unless you yourself lay down what is to be so called. But you will find it difficult to hit upon such a convention; at least any that satisfies you.
>
> *(Wittgenstein 1958, 42)*

The inexact results – which are however useful to the peasant, for instance – are often considered 'mistakes' by school mathematics. However, Foucault writes, 'but perhaps there are no errors in the strict sense of the term, for error can only emerge and be identified within a well-defined process' (Foucault 1972,

222). The language games of *cubação*[1] of the land (Knijnik, 2006, 2007), when examined in the contingency of the peasant forms of life of the Landless People, do not present any error *stricto sensu*. Since they are useful in decision-making for cultivation practices, and also easy to use, the peasants do not disqualify these local knowledges.

The second sentence opening the chapter also refers to this same kind of 'inexactness': 'I live very close to my work. It takes 10 minutes by bus to get there'. This was a statement made by Maria, a 52-year-old woman, who attended an adult education project developed by a university in the capital of the southernmost state of Brazil. The course was addressed to people who had only finished the first years of obligatory schooling in the country. As the initial activity of the course, her teacher administered a questionnaire to all students. The teacher explained when she was interviewed that the organisation of this pedagogical activity had two objectives. The first was to obtain information about each member of the group, such as age, profession and workplace. This would tell her more about the students and their forms of life. 'Here is something that I value greatly. If one knows the students, one can plan activities related to their reality: it is very important to bring the reality of the students to the classroom'. The second objective was eminently mathematical: to begin the study of tables and graphs. Based on her knowledge of the students' everyday practices, the teacher considered these content areas to be relevant for their development of numeracy.

After the students completed the questionnaire, the 15 students were divided into five groups. The set of questions was also divided into five, and each group was asked to be responsible for summarising one of the subsets of the questionnaire. The students were asked to make tables and graphs with the data obtained from the answers. One of the questions in the questionnaire was drafted as follows: 'How far is it from your home to your work?' Maria used time – 10 minutes – to express the displacement from her home to her workplace. Following Bauman (2000), we can say that the meaning given by Maria to the time-space relation is not anything new. In the sociologist's own words:

> When I was a child (and that happened in another time and another space) it was not uncommon to hear the question 'How far is it from here to there?' answered by 'About an hour, or a bit less, if you walk briskly.' In the time more ancient yet than my childhood years, the more usual answer, I suppose, would have been, 'If you start now, you will be there about noon', or, 'Better start now, if you want to be there before dusk.' Nowadays, you may hear on occasion similar answers. But it will be normally preceded by a request to be more specific: 'Do you have a car? Or do you mean on foot?' 'Far' and 'long', just like 'near' and 'soon', used to mean nearly the same: just how much, or how little effort would it take for a human being to span a certain distance – be it by walking, by ploughing or harvesting.
>
> *(Bauman 2000, 171)*

Bauman suggests that, beginning with modernity, time began to occupy a privileged place. 'History of time began with modernity. Indeed, modernity is, apart from anything else, perhaps even more than anything else, history of time: the time when time has history' (Bauman 2000, 172). More incisively, in the contemporary context, this privileged place is taken up in forms of life of large urban centres. In the globalised world in which we live, the issue of mobility of individuals and also of merchandise, is especially noticeable (Bauman 2006). His analysis refers to two notions: solid modernity and liquid modernity.

In *Liquid Modernity* (Bauman 2006), the sociologist uses the term 'liquidity' to characterise the state of present-day society, and establishes an analogy between the latter and the 'solidity' that would mark the previous period. In this metaphor, the author argues that because liquids do not stay in a fixed and stable form, they do not capture space nor imprison time, moving faster than the solids. Because of this constant possibility of change, liquids can be associated with lightness. 'These are reasons to consider "fluidity" or "liquidity" as fitting metaphors when we wish to grasp the nature of the present, in many ways novel, phase in the history of modernity' (Bauman 2006, 2).

Solid modernity, according to Bauman (2006), seeks to form solids with the marks of stability, fixity and safety, which might make the world more predictable and, thus, better regulated. This search for order, balance and also for regulation, neglects contingency, variety or instability. Thus, the different characteristics of the liquids and solids lead Bauman to associate the marks of diffusion and capillarity with fluid modernity, while solid modernity is then configured as heavy, condensed and systemic. His discussion (Bauman 2000) emphasises that in solid modernity, it is essential to conquer the space; its possession is even a modern obsession. Wealth and power are geographical concepts tied to place and land and property.

The above discussion brings up questions that directly concern numeracy as a social practice in its relations with the school processes of teaching and learning mathematics. Considering numeracy as a social practice implies accepting that what we usually call 'mathematics education' is best expressed by the term 'school mathematics education', namely, that offered by the school system based on Western obligatory schooling. This qualifier attached to mathematics education converges with the positions of Valero (2009) and Knijnik (2010). According to these authors, the processes involving learning and teaching mathematics are developed in school and non-school spaces.

Indeed, Valero (2009, 17) understands mathematics education as a 'network of social practices that operate in different spaces and at different levels, not limited to classrooms and the school context'. Building on this understanding, Knijnik (2010), from her ethnomathematics formulations, anchored by the later ideas of Wittgenstein, conceives mathematics education as the processes of teaching and learning the mathematics language games of different forms of life, including the school form of life. This implies accepting that mathematics language games are also practised in non-school forms of life. Historically, ethnomathematics was the field of mathematics education that extended the analysis of mathematics language games beyond school, especially those practised by adults in their work and domestic activities (Ferreira 1993).

Numeracy and the field of ethnomathematics

In recent decades, migratory movements have increased worldwide, leading to many immigrant children being schooled in global North countries, such as those belonging to the European Union and the United States. Most of their families can be considered functionally illiterate, which makes it even more difficult to find jobs that can allow them to have a decent life. The field of education and particularly mathematics education gained awareness of the need to pay attention to this sociocultural phenomenon that has concrete repercussions in schooling and broader social contexts (Abreu 2001; Civil 2007). One of the most significant trends in mathematics education that expresses this awareness is ethnomathematics.

Since its beginning, ethnomathematics was linked to numeracy, as conceived by Johnston and Yasukawa (2001). One of its sources was the movement called Popular Education developed in Africa and Latin America in the 1960s. The ideas of Paulo Freire in Brazil and in other global South countries, indicating the politicity[2] of education, its non-neutrality, and its role in constructing a more just and egalitarian society, reached the sphere of mathematics education (Frankenstein 1987; Powell and Frankenstein 1992; D'Ambrosio 1997). Ubiratan D'Ambrosio – the Brazilian educator internationally acknowledged as the one who coined the term 'ethnomathematics' – received as one of his intellectual influences Freire's ideas on education (Freire, D'Ambrosio and Mendonça 1997; Higginson 1997). Furthermore, D'Ambrosio's own previous professional trajectory in the south of the USA greatly influenced his work. (D'Ambrosio 1993; Ascher and D'Ambrosio 1994; Chassot and Knijnik 1997).

One of the main convergences between ethnomathematics and Freirean thinking regards the value attributed to people's culture. As Freire pointed out already in his first works, the ways people produce meanings, how they understand the world and live their daily lives, are taken as major, even core, elements of the educational process. Using Wittgenstein's ideas, Freire's position can be interpreted as saying that mathematical language games practised in non-school forms of life must be part of the school curriculum. Nevertheless, it is necessary to be careful not to produce an exacerbated relativism, a naïve view of the potential of such people's knowledges in the pedagogical process, which might lead to a glorification of popular knowledges with the consequent ghettoisation of the subordinated groups (Grignon 1992). As with other fields of knowledge, since its beginning ethnomathematics has been a vast, heterogeneous area of study, rendering it impossible to establish generalisations about its theoretical-methodological contributions. Even so, the studies in this field can be understood as comprising works aimed at identifying, recognising and valuing the mathematical language games produced in different forms of life (Barton 2004).

As written elsewhere (Knijnik 2010), one of the main risks of these works is the 'domestication' of the indigenous knowledge, i.e. the subordination of this knowledge to school mathematics and the curriculum logic to which it is attuned. Wittgenstein's ideas in his later works show very clearly how the displacement of a language game from one form of life to another is an unsuccessful operation. Another risk of ethnomathematics refers to research work which mainly highlights the mathematical contents of the 'indigenous' practices connected to 'our' mathematics. This kind of research reinforces the hegemonic mainstream mathematics, precisely what ethnomathematics would intend to problematise. Dowling (1991) very clearly discusses this point. Other important critical analyses of ethnomathematics have been made since this field emerged; it is worth mentioning the works of Skovsmose and Vithal (1997), Rowlands and Carson (2002) and, more recently, the paper published by Pais (2011) which extends the analytical criticisms about ethnomathematics.

Knijnik's ethnomathematics perspective, which provides theoretical support for this chapter, allows us to discuss some of Dowling's critiques. This perspective is considered as a theoretical toolbox built upon the notions of Wittgenstein's later work and of Michel Foucault's thinking. It enables 'examining different mathematical language games and their family resemblances, and the discourses that constitute academic and school mathematics, analysing their effects of truth' (Knijnik 2012, 91). Following Wittgenstein's and Foucault's philosophical conceptions, we can establish a dialogue with Dowling's critiques of ethnomathematics. The 'plural monoglossism' mentioned by the English researcher can be explained in other terms; it is not a matter of saying that ethnomathematics considers society as heteroglossic but each form of life monoglossic. First, because forms of life are not isolated at all (especially in the globalised world in which we live), this destroys any possibility of 'pure' monoglossism. Moreover, it becomes clear how Knijnik's ethnomathematics perspective goes against the existence of a single (and universal) rationality, i.e. modern thinking. Finally, it should be mentioned that the reasoning developed in this section offered elements to show the consistency of articulating Wittgenstein's and Foucault's philosophical ideas: their convergent anti-foundationalism and anti-essentialism attitudes and their common understanding about language.

If we look at ethnomathematics as it is being signified today, we can state that almost 40 years since it emerged, it continues to be interested in discussing the politicity of dominant knowledge practised in school. This politicity can be considered in two dimensions. In the first, dominant knowledge maintains its position through compartmentalisation, placing the knowledge of the world in incommunicable drawers, making us believe that it is 'natural' for school to be organised by disciplines, that school time and space are distributed among mathematics lessons, history lessons, language lessons, science lessons etc. We can then ask ourselves: Is this the only possible way of organising the school institution?

The second dimension regarding the politicity of dominant knowledge refers to the subtle manoeuvre that hides and marginalises given contents, given knowledges, forbidding them in the school curriculum. Everything seems 'natural' to us, 'as it has always been'. So one must ask: Is there a way to build other schooling processes, 'another' school that would include other contents, and not only those that usually circulate in the school curriculum? We have been shaped in such a way, normalised by what is usually called 'knowledge accumulated by humankind', that we do not even dare imagine that what we call 'knowledge accumulated by humankind' is only a small part, a very specific part, of the much broader and more diversified ensemble of what has been produced throughout history. These questions lead us to think about new possibilities of conceiving school mathematics education.

What to do on Monday mornings?

The title of this section was formulated by Paul Willis (1977) in another context that was not mathematics education. However, it has been the source of inspiration for our work as researchers and teachers in the field of mathematics education. Our thinking is not only linked to theoretical issues such as those that we have presented in this chapter. One of the lessons that we have learned in our work in teacher education and in the follow-up of traineeships in primary and secondary schools and in adult education projects concerns the 'Monday morning' imperative. That is when we enter the classroom to carry out our everyday work. We have asked ourselves how to make our teaching and our research compatible: How does this kind of analytic exercise contribute to the 'Monday morning' syndrome?

The first answer is to explore the possible contributions offered by this kind of analytic exercise. This will give us elements to incorporate into our multiple school mathematics education, reflections about the place occupied in contemporaneous society by mathematics, by its connections to the broader social world, and by its uses in other fields of knowledge. As we discussed elsewhere (Knijnik 2014), the mathematics produced by the mathematicians has taken a central place in contemporary technical-scientific production. This ultimately has repercussions on school mathematics education, which is positioned as the most important school discipline. For example, in Brazil this means that in the school schedule the greater number of hours is addressed to school mathematics. This highest place is also responsible for dropouts and failing grades.

In this scenario, school gives special attention to the acquisition of mathematics language games whose rules follow the formalism and abstractions characteristic of academic mathematics. These are the games that, over time, have shaped what we call school mathematics. We have argued for the importance of expanding our students' repertoire of mathematical language games by including in this repertoire mathematical language games practised in non-school forms of life. The mathematical language game practised by the landless peasant to calculate the portion of land to be cultivated and the language game used by Dona Maria to express the distance from her house to her work place have the marks of contingency. They are considered valid in their respective forms of life. They are different from the school mathematics language games that have an outstanding transcendence, a universal character.

However, there is an important point to highlight: in school mathematics education the expansion of the mathematical language games repertoire, to include those practised in non-school forms of life, must necessarily be accompanied by pedagogical work. In this work the different rationalities that mark the distinct forms of life must be discussed, to demarcate the rules that conform to each of their logics. It is important to reflect on questions such as those we discuss here, going to their roots. We believe that, as Foucault (2008) taught, this is a fruitful path for the emergence of counter-conduct movements. Like them, doing so we may be different from what we are today as individuals, as teachers. Maybe these counter-conduct movements can somehow become 'lines of flight' for another world of possibilities (Guattari 2016) that favour changing what is established and inventing other 'Monday mornings'.

Notes

1 Flemming et al (2005: 41) define *cubacao* as 'the problem of measurement of land using diverse shapes'; see Knijnik 1996.
2 Politicity is understood as the search for a new sense of life and politics, which can potentially renew the idea of change and forms of social action.

References

Abreu, G. (2001) "O papel mediador da cultura na aprendizagem da matemática: a perspectiva de Vygotsky" *Educação, sociedade e cultura*, 13: 105–117.
Ascher, M. and D'Ambrosio, U. (1994) "Ethnomathematics: a dialogue" *For the Learning of Mathematics*, 14: 36–43.
Barton, B. (2004) "Dando sentido à etnomatemática: etnomatemática fazendo sentido". In Ribeiro, J. P. M., Domite, M. C. S. and Ferreira, R. (eds), *Etnomatemática: papel, valor e significado* São Paulo: Zouk.
Bauman, Z. (2000) "Time & Society" *Sage*, 9: 171–185.
Bauman, Z. (2006) *Liquid modernity* Cambridge: Polity Press.
Chassot, A. and Knijnik, G. (1997) "Conversando com Ubiratan D'Ambrosio" *Episteme: filosofia e história das ciências em revista*, 2: 96–110.
Civil, M. (2007) "Building on community knowledge: An avenue to equity in mathematics education". In Nasir, N. and Cobb, P. (eds), *Improving access to mathematics: Diversity and equity in the classroom* New York: Teachers College Press.
D'Ambrosio, U. (1993) "Etnomatemática: Um programa" *Educação matemática em revista*, 1: 5–11.

D'Ambrosio, U. (1997) "Where does ethnomathematics stand nowadays" *For the Learning of Mathematics*, 17: 13–17.

Díaz, E. (2005) *La filosofía de Michel Foucault* Buenos Aires: Biblos.

Dowling, P. (1991) "The contextualizing of mathematics: Towards a theoretical map". In Harris, M. (ed), *Schools, mathematics and work* Basingstoke: The Falmer Press.

Ferreira, E. S. (1993) "Cidadania e educação matemática" *Educação matemática em revista*, 1: 12–18.

Foucault, M. (1972) *The archaeology of knowledge: And the discourse on language* New York: Pantheon Books.

Foucault, M. (2008) *Segurança, território, população* São Paulo: Martins Fontes.

Frankenstein, M. (1987) "Educação matemática crítica: uma aplicação da epistemologia de Paulo Freire". In Bicudo, M. A. V. (ed), *Educação matemática* São Paulo: Moraes.

Freire, P., D'Ambrosio, U., and Mendonça, M. C. (1997) "A conversation with Paulo Freire" *For the Learning of Mathematics*, 17: 7–10.

Glock, H. J. (1996) *A Wittgenstein dictionary* Oxford: Blackwell.

Grignon, C. (1992) "A escola e as culturas populares: pedagogias legitimistas e pedagogias relativistas" *Teoria & educação*, 5: 50–54.

Guattari, F. (2016) *Lines of flight: For another world of possibilities* London: Bloomsbury.

Higginson, W. (1997) "Freire, D'Ambrosio, opresion, empowerment and mathematics: Background notes to an interview" *For the Learning of Mathematics*, 17, 3–4.

Johnston, B. and Yasukawa, K. (2001) "Numeracy: Negotiating the world through mathematics". In Atweh, B., Forgasz, H. and Nebres, B. (eds), *Sociocultural research on mathematics education: An international perspective* Mohwah, NJ: Lawrence Erlbaum.

Knijnik, G. (2006) *Educação matemática, culturas e conhecimento na luta pela terra* Santa Cruz do Sul: Edunisc.

Knijnik, G. (2000). "Cultural diversity, landless people and political struggles". In Ahmed, A., Williams, H. and Kraemer, J.M. (eds) *Cultural Diversity in Mathematics (Education)* CIEMAEM 51. Chichester: Horwood Publishing.

Knijnik, G. (2010) "Educação (matemática) do campo e movimentos sociais". In Dalben, A., Diniz, J., Leal, L. and Santos, L. (eds), *Convergências e tensões no campo da formação e do trabalho docente* Belo Horizonte: Autêntica.

Knijnik, G. (2012) "Differentially positioned language games: Ethnomathematics from a philosophical perspective" *Educational Studies in Mathematics*, 80: 87–100.

Knijnik, G. (2014) "Juegos de lenguaje matemáticos de distintas formas de vida: contribuciones de Wittgenstein y Foucault para pensar la educación matemática" *Educación matemática*, 25: 146–161.

Knijnik, G. and Wanderer, F. (2015) "Mathematics education in Brazilian rural areas: An analysis of the public policy and the Landless Movement pedagogy" *Open Review of Educational Research*, 2: 143–154.

Pais, A. (2011) "Criticisms and contradictions of ethnomathematics" *Educational Studies in Mathematics*, 76: 209–230.

Powell, A. and Frankenstein, M. (1992) "Toward liberatory mathematics: Paulo Freire's epistemology and ethnomathematics". In McLaren, P. and Lankshear, C. (eds), *Conscientization and oppression* London: Routledge.

Rowlands, S. and Carson, R. (2002) "Where would formal, academic mathematics stand in a curriculum informed by ethnomathematics? A critical review of ethnomathematics" *Educational Studies in Mathematics*, 50: 79–102.

Skovsmose, O. and Vithal, R. (1997) "The end of innocence: A critique of 'ethnomathematics'" *Educational Studies in Mathematics*, 34: 131–157.

Valero, P. (2009) "Mathematics education as a network of social practices" presented at VI Congress of the European Society for Research in Mathematics Education, Institute Français de l'Éducation, Lyon.

Walkerdine, V. (1995) "O raciocínio em tempos pós-modernos" *Educação e realidade*, 20: 207–226.

Willis, P. (1977) *Learning to labor: How working class kids get working class jobs* New York: Columbia University Press.

Wittgenstein, L. (1958) *Philosophical investigations* Oxford: Blackwell.

8

'LIMITS OF THE LOCAL' IN THEORISING NUMERACY AS SOCIAL PRACTICE

A case study of mathematics education in Palestine

Jehad Alshwaikh and Keiko Yasukawa

Introduction

Mathematics education research experienced a 'social turn' in the late 1980s (Lerman 2000) which led to mathematics education researchers looking beyond traditional cognitive theories to social theories for explanations about how people learnt mathematics (or not). Researchers started to seek understandings about mathematics learning not only by interrogating the cognitive skills or processes of the individual learner, but by asking questions about the social, cultural and political contexts in which the learning was taking place. Perspectives on mathematics and numeracy as a social practice emerged during this period. In mathematics education, where what Baker (1998) calls the 'autonomous' model of mathematics had been and arguably still is so prevalent, the fact that there was a 'social turn' at all is remarkable in itself. However, the 'social turn' has given rise to important insights about what people 'do' with mathematics (including how they produce new mathematical practices) and how the 'doing' is highly contingent on where, with whom and for what purpose the 'doing' is situated. A distinctive feature of the social practice perspectives, both in numeracy and literacy, is the foregrounding and legitimation of the local practices that afford context-specific meaning to those who are involved in the activities in which literacy and numeracy are embedded. This is presented as an alternative model of numeracy by Baker (1998) to the autonomous view of numeracy and mathematics as a singular, universal knowledge system.

The social practice perspectives of numeracy have enriched mathematics education research by uncovering a number of numeracy practices that, until then, had been overlooked, both by the 'doers' of mathematics and mathematics education researchers. However, as the social practice perspectives gained momentum, considered critiques also emerged, including from Brandt and Clinton (2002) in their paper 'Limits of the local' which questioned the extent to which literacy could be theorised solely from an examination of the social contexts in which these practices occur. They argued that what the social practice perspectives can offer could be enriched by understanding that local practices also have 'transcontextualised and transcontextualising' aspects (Brandt and Clinton 2002, 337); in other words, not all aspects of a local practice are locally produced or contained. Drawing on ideas from actor-network theory (Latour 1987), Brandt and Clinton argue that literacy is an 'actant' in local practices: 'it participates in social practices in the form of objects and technologies whose meanings are not usually created or exhausted by the locales in which they are taken up'. There are, they argue, aspects of literacy that have the ability to 'travel, integrate and endure' (Brandt and Clinton 2002, 337); that is, material objects must also be considered as actors in literacy practices, serving as 'comrades, colleagues, partners, accomplices or associates in the weaving of social life' (Latour 1996, in Brandt and Clinton 2002, 348).

In this chapter we consider the meaning of 'limits of the local' in relation to the notion of numeracy as a social practice through a case study of mathematics teaching and learning in Palestine. We find that in numeracy practices too, there are material objects that travel, integrate and endure across contexts, and which find themselves endowing meaning, in this case, in the context of Palestinian mathematics classrooms. This case study evolved through a series of email 'conversations' between the authors: Jehad in Palestine, who was the source of information about mathematics education in Palestine, and Keiko in Australia, who as the critical friend initiated the conversation. As the conversation continued, and Jehad spoke about his research interest in the role of textbooks in Palestinian mathematics classrooms, it transpired that the 'textbook' was a critical travelling actant that integrated particular 'autonomous' aspects of mathematics teaching and learning into the Palestinian school contexts and in an enduring way. We therefore argue that it is an example of what Latour calls an 'immutable mobile' (Latour 1990), a material object that travels largely intact across diverse cultural, linguistic and political contexts. However, rather than adopting a deterministic or fatalistic understanding of the practices brought about through these textbooks, we explore a possibility that Brandt and Clinton (2002) suggest in relation to literacy, that is, the possibility to 'rehabilitate' certain aspects of the 'autonomous' models of numeracy. This possibility is explored in this chapter, aided by Jehad's research on the role of language in mathematics meaning making.

The linguistic turn in mathematics education research

In addition to the 'social turn' in mathematics education, there was also a 'turn to language' (Morgan 2006, 219) in mathematics education which drew attention to the role of language in meaning-making in mathematical activities, and the study of the different discourses of mathematics and mathematics teaching and learning. While different motivations can be found in the literature for researching the role of language in mathematics teaching and learning, many are connected to a similar social justice concern about the hegemony of the 'autonomous model' of mathematics in educational contexts that motivated Baker (1998) and many others who have contributed to the social turn in mathematics education research (see for example, Barwell et al. 2016; Halai and Clarkson 2016).

A number of mathematics education researchers, including Jehad, have been influenced by the works of Halliday's 'language-based theory of learning', founded on the view of human learning as 'a process of making meaning – a semiotic process' (Halliday 1993, 93). Halliday talks about language not as an object of study (except when one is studying linguistics), but as 'the essential condition of knowing, the process by which experience becomes knowledge' (Halliday 1993, 94). His functional theory of language, Systemic Functional Linguistics (SFL), thus focuses on how language functions as a meaning-making system, a perspective that resonates with the Vygotskian view of language as a mediating resource for learning. He explains how language performs three functions: *ideational*, *interpersonal* and *textual* (Halliday 1978). By ideational, Halliday means the function of the language to represent the speaker's/writer's experience of phenomena in their world: it is 'language as "about something"' (Halliday 1993: 112). The interpersonal function serves to express the relations among the speaker/writer and the listeners/readers, and the textual function enables the 'something' that is being expressed to be organised in such a way that it is coherent. The ideational, interpersonal and textual functions of language thus work together to enable meaning-making to occur. Learning language, from this perspective, is learning how to make meaning of one's world, including the mathematics teaching and learning environment.

In his book *Language as Social Semiotic*, Halliday (1978) writes about the sociolinguistic aspects of mathematics education, drawing attention to features of the *register* of mathematics: 'the meanings that belong to the language of mathematics' (the mathematical use of natural language: that is, not mathematics itself). Since Halliday's early writings, other social semioticians such as Iedema (2003), and Kress and Van Leeuwen (1996) extended the focus on language as a meaning-making system to multimodal texts – that is the incorporation of other modes of communication such as images. Gunther Kress (2010) refers to multimodality as the use of multiple modes or ways of representation and communication. Mathematical discourse is a multimodal discourse where different modes, including images, symbols, graphs and gestures are deployed to make meaning.

It is precisely the affordances inherent in the multimodality of mathematical discourses, as Jehad will explain, that may lead to a 'rehabilitation' of the autonomous, hegemonic aspects of mathematics education. We postulate the potential of exploiting the multimodal affordances in classroom discourses to disrupt the discourse that textbooks can so easily establish in the classrooms.

Exploring the 'limits of the local' in mathematics teaching and learning in Palestine

Mathematics travels the world

Keiko: Would you say a little bit about your own history in mathematics education in Palestine, prior to your decision to research in this area?

Jehad: Mathematics in Palestine, and probably in many other contexts/ countries around the world, is seen as a hard topic that is made for 'nerds' and smart people. I never felt that way to be honest with you, although I was good at mathematics in school. The situation was different at the university! There is a funny, but also serious story for me that is attached to the word 'function'. I was born and raised in a refugee camp in Gaza city where I studied the Egyptian textbooks including for mathematics. Palestinians in the West Bank, on the other hand, studied the Jordanian textbooks. The Egyptian texts used the word 'daleh' (دالة) as the translation for 'function'. The Jordanian textbooks used the word 'iqteran' (اقتران) for function. It took me a year and very low performance in maths at the university to find out that the 'function' in the English textbook used there, and the word 'iqteran' that the lecturer used was the same word as 'daleh'. I felt awful about that and often asked myself: did it not occur to the lecturer to mention that there were different translations of the word 'function'; especially given the students came from different contexts?

Keiko: I can imagine the confusion! So, are you saying that at university, the lecturer was referring to the term used in Jordanian textbooks – or at least the Jordanian term for 'function'?

Jehad: Yes, he used the Jordanian term (iqteran). He was using an English textbook (an American one, I guess). Most of the faculties of science (including mathematics) in the Palestinian universities use English textbooks.

I want to refer to the disconnect between school mathematics and academic or university mathematics (with one common feature – that mathematics remains very removed from the daily life and experience, as Fasheh (1996) had observed – except in the use of calculations). Mathematics topics are also presented as separated islands; linear algebra has nothing to do with graph theory, for instance. Euclideian geometry has no place in many university mathematics courses even though we, the graduates, are expected to teach it in schools when we go to teach there.

Keiko: This is an interesting reflection. I majored in mathematics at university and it was only in my Honours year when I studied a subject called category theory and another called sheaf theory where the connections were made between all the 'islands' – though the connections were at such an abstract level that none of it really made a lot of sense to me.

Jehad: Thanks for sharing. It is interesting to hear that from you especially that you were a pure mathematician, as I know. Are you saying that even if we teach or learn such connections, it still is not enough as long as the connections are abstract? In other words, we add another 'separated island' to the other 'mathematical islands' if we still insist in teaching mathematics in the same way that is abstract and context-free. Maybe this is true.

Later, after my graduation, I assisted in developing general guidelines for the mathematics curriculum for the newly proposed Palestinian curriculum when the Palestinian Ministry of Education was founded after the political agreement between the PLO (Palestinian Liberation Organisation) and Israel in 1994. Our main activity was to look at different textbooks around the world (American and British for example) and suggest the new Palestinian curriculum. In other words, our suggestions were not based on our own needs as Palestinians, and we were limited to 'borrowing' other textbooks. On reflection, that endeavour was informed by the notion that mathematics was a 'global language' and not related to social and political practices although the whole issue of creating the new Palestinian curriculum was mainly informed by the political change!

Keiko: What a great example of that 'universality' of the 'autonomous model' of mathematics. You mention American and British textbooks. Did you look only at textbooks from the 'first world' countries, and countries where English was the dominant language? What guided your selection of countries from which you sourced textbooks?

Jehad: Yes. I still, however, remember looking at European texts such as the Dutch ones. We also looked at some Arabic textbooks from the Arab countries such as Jordan and Egypt. However, the Jordanian textbooks influenced the actual or the produced textbooks that continue to be heavily used in Palestinian schools.

Keiko: How different were the textbooks from the English-speaking countries, European countries, Egypt and Jordan? And what made you settle on the textbook used in Jordan as the model?

Jehad: Interesting question! If I remember correctly, there were not many differences in terms of the 'mathematical content' – maybe some textbooks included more lessons or concepts. I think that we looked first at the main mathematical themes or big ideas of the content in the many textbooks we reviewed and then decided what themes to include in the new suggested textbooks, as well as how much to include of each theme (quantity wise). Choosing the Jordanian model is not a surprise when I look back. One reason is due to the political and historical connections between the West Bank (where the focus of the Palestinian Authority was and still is) and Jordan. The second reason is the fact that the Palestinian Ministry of Education, newly established then, had used the Jordanian textbooks in the Palestinian schools before producing the first Palestinian textbooks.

Jehad's examination of mathematics textbooks from different parts of the world attest to the strengths and the ubiquity of the belief that mathematics is 'the same' in different educational contexts. The belief that a textbook from one context could be unproblematically adopted and adapted in another context resonates with Brandt and Clinton's (2002) observation in relation to literacy, that some aspects of literacy practice travels across contexts. But what travels also brings with it and endows particular meaning in its destination, for example, the privileging of abstract mathematical knowledge over everyday mathematical practices.

From the global to the local

Keiko: I noticed in some of your earlier publications that you chose textbooks as a lens to examine discourses in Palestinian mathematics classrooms. Is this correct? Can you explain why you decided to investigate mathematics education discourses?

Jehad: I would agree with that, though my investigations are not limited to that. As I mentioned earlier, I assisted in developing the mathematics curriculum at the Palestinian Curriculum Development Centre. The political situation in Palestine played a role here together with my PhD study in communication and representation in mathematics discourse. They drew my attention to the social aspects of teaching and learning mathematics – an issue that is so neglected in Palestine except among some researchers and educators.

             I started to see the connection between mathematics as an academic topic that I am interested in, and the socio-political situation that I live in. I started to see the impact of the language use and the other modes or means of communication in mathematics textbooks (visual and gestural representations for example) and how they worked to reproduce the dominant view of mathematics as an abstract and formal topic.

Keiko: So it sounds like you had a bit of a 'light bulb moment'. Can you remember when and why you started to be drawn to the social aspect of teaching and learning mathematics? I am guessing that there would be an equal number of maths teachers – or probably many more, who could be working in Palestine and other complex sociopolitical contexts who, for various reasons, would make no connections between what happens in the maths classroom and what is happening immediately outside the school walls.

Jehad: Your guess is right and I would agree that that case is valid for many teachers around the world. In my case I had to wait until I did my PhD to be introduced to the social aspects of maths education. Probably it was the process of connecting different threads and starting to see the social and political dimensions of maths. However, I still remember my amazement in reading about that as I started to focus on language and its connection to mathematics discourse. Candia Morgan (with whom I worked) had been writing about this too. So I wanted to look closely at mathematics textbooks and analyse the kind of discourse that is used to encourage this dominant view of mathematics. I wanted to contribute to changing this view or at least to offer different views.

Keiko: What you are suggesting about the reproduction of the dominant view of mathematics is interesting. Whilst I wasn't thinking about the sociopolitical implications of this when I was a university mathematics student, I found it remarkable (and useful) that I could consult both textbooks written in Japanese and English, and find the presentations and explanations of mathematical ideas so similar. As part of the requirement of my doctorate, I even managed to translate (well enough) a small amount of mathematics text written by the Bourbakí group[1] from French to English without ever having studied French because the genre was so familiar to me. If anyone at the time questioned the universality of mathematics as a knowledge system, I would have questioned if they had any idea what they were talking about.

But you mentioned how, using multimodal analysis, you found the visual and gestural representations reinforced the reproductive power of textbooks. Can you elaborate on this?

Jehad: In multimodal analysis of mathematical texts/discourse, we look at the different modes together and attempt to 'read' and interpret the communication and representation in a way that enables us to say something about the nature of the mathematical activities as well as the social relations represented in the text.

I worked with Candia Morgan and focused on diagrams (as visual communication) mainly in geometry. It is very interesting to observe how diagrams as visual representations had been relegated, historically, from being an integral part (if not the essential part) of mathematical texts to a neglected one that should be removed when published. However, 'we' still can read the mathematical activities and deduce something about the nature of such activities as well as the associated social relations – for example, the relationship between the creator/author of the diagram and the viewer/user of it.

Reading mathematical activities required looking at the nature of the diagrams represented in the mathematical text. One of the interesting questions I was thinking about during my PhD study is why we use dotted lines in drawing diagrams in mathematics. I make a distinction between narrative and conceptual diagrams. In the former, one can follow a 'timeline of action' – for example, an arrow suggesting a shift in position (rotation of a triangle by 45°), or dotted lines indicating that part of a shape has been divided by a perpendicular line. In a conceptual diagram, no such timeline can be traced – the diagram simply describes properties of a mathematical object, for example a geometric shape. I refer to the difference between the two kinds of diagrams in terms of a *temporality* feature. (More details in Alshwaikh (2011) in which I suggest specific characteristics or indicators for such difference.) Conceptual diagrams are present more often in (Palestinian) mathematical texts and other contexts as well.

Keiko: When you talk about a 'timeline', are you referring to a timeline of sense-making in which the viewer/reader is engaged? For example, the dotted line in a diagram such as the one in Figure 8.1, about the concept of *reflective symmetry,* guides the viewer to notice particular aspects of the two halves created by the dotted line – that is, that they are a *reflection* of each other.

And would an example of a conceptual diagram be something like the following which 'defines' what a right-angled triangle is by showing one of the angles as being 90° (Figure 8.2).

Jehad: I argue that narrative diagrams portray mathematical activities as involving humans, while conceptual diagrams present atemporal mathematical objects or relationships, removing human actions. There are specific indicators one can actually see in narrative diagrams such as dotted lines. Thus in your first example, the dotted line was added to the original triangle (we know that because of the common practice in doing mathematics to distinguish that line of action as dotted, not solid as the rest of the sides). The word 'added' that I just used suggests time and action. In contrast, in the right-angled triangle example, as you put it correctly, the diagram presents that kind of triangle as an object in a timeless or enduring way. Thus conceptual diagrams may reinforce students' views that mathematical knowledge is a 'given', not dynamic with some human agency involved, and therefore fails to serve

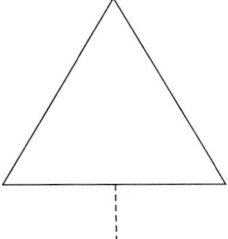

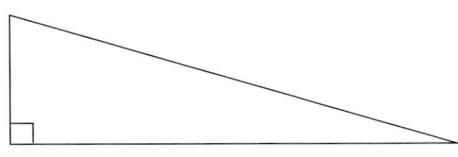

FIGURE 8.1 Dotted line indicating reflective symmetry

FIGURE 8.2 Diagram indicating a right-angled triangle

the interpersonal function that is needed for the diagram to be a meaning-making resource for the students.

Keiko: Interesting to see how ideologies are embedded in mathematical diagrams!

You have also used Kress and Van Leeuwen's (2006) idea that the interpersonal function of visual texts is realised by contact – how the object of the visual is labelled, the social distance – whether the diagram is a polished disembodied 'product' or a work in progress, and modality – the degree of certainty or tentativeness in the diagram. Reflecting on my own experiences with mathematics texts, I can reasonably confidently say that most of the diagrams in the texts appeared to present the 'truth', which I was either smart enough to 'see', or too stupid (lacking in a mathematical bent) to appreciate. It's interesting to consider the extent to which the diagrams scaffolded my learning (or not), and to what extent I felt sometimes that the diagram was 'speaking' to me, or only to a much 'cleverer' mathematician.

It is in the mathematics classrooms that the mathematics textbooks that are imbued with values such as the universality of particular forms of mathematical knowledge face the cultures and practices of the locale. However, in Jehad's account, the semiotic resources utilised in the textbooks work largely to protect and reinforce the dominant view of mathematics from being challenged by any local interpretations. Consequently, the students cannot access the meaning-making potential of the texts.

New possibilities through a social semiotic perspective

Keiko: Given that mathematical discourses are multimodal, do you think that there are ways of using texts, visuals, talk and gestures in ways that enable students to make meaning of mathematical ideas in their terms, or at least in a way that is connected to their lives?

Jehad: I totally agree with you. And this is probably what was of interest to
           me – to bring the social and the political into the classroom through
           the mathematics discourse. The connection is very interesting.
           You reminded me of something I wrote to myself about my little
           journey in Ramallah in West Bank when I was teaching a maths
           education course and I insisted on using a newly launched book at
           that time in 2012 *Equity in Mathematics Discourse*. Can I say it here?

           I was wondering what I'm doing with my students from
           these far villages asking them to read in English about
           mathematics education? What does it mean to read about
           equity in mathematics discourse? What are, if any, the benefits
           of teaching Geogebra [a calculator with graphing function]
           to pupils/learners in those remote villages who know mainly
           road number 443, settlements, Jews, checkpoints, abu Mazen
           and Fateh?[2]

           I'm not sure how relevant is this reflection here, but I want to
           refer to the fact that the social view of maths enabled me to see
           such connection between the social and the political and maths
           education even with a painful and sceptical tone/mood.
               I think that currently there is a very strong relationship between
           what is offered in school maths teaching and learning and the way
           that maths is represented in textbooks. Maths is mostly presented
           as formulae and procedures that students have to follow in order
           to solve problems correctly. Textbooks in Palestinian schools are
           centralised – only one textbook for each grade. Teachers mostly
           follow textbooks as they are, for different reasons – worried about
           their jobs, not feeling that they have the agency to change or
           influence them, have no time to plan carefully, etc. This is a key
           point here – obedience and discipline takes you a long way – an
           observation you made during our conversation.
               My main observation about the textbook so far is that the
           mathematical theory, formula, or law will be presented at the
           beginning of the lesson followed by examples, an activity and lastly
           the problems. This is the most probable thing Palestinian teachers
           would do in a maths classroom: they will start by writing the
           theorem or the law, solve one or two examples (mostly the same
           ones as the textbook), ignore the activity suggested in the textbook,
           ask the students to solve problems from the textbook on their own,
           invite some students to solve those questions and finally announce
           the numbers of problems to be solved as a homework.

Keiko: From what you have just described, it sounds like the textbook is
           not only playing a large role in determining what constitutes the

mathematical knowledge that students need to learn, but also how the teachers and students are to make meaning of this knowledge.

Have you observed any difficulties or resistance from the students to the ways of working with mathematics that are promoted in the textbooks? If so, can you give one or two examples? Do you have any insights into the source of resistance or difficulties?

Jehad: The common question that teachers would mention and feel the challenge of is that students ask what is the point of learning maths? How is it connected to life? I guess that teachers themselves have the same question. The basic message from the students for me is 'why do you teach this subject in this abstract way?', as if they are saying: we [students] do not understand what you [teachers or textbooks] are talking about. For me, it is the way that maths is offered and communicated in textbooks: one view of maths and one way of doing 'things'. The dominant view is that maths is abstract and symbolic and that there is only *one right* way to do it. The *quicker* you are in solving the problems, the smarter you are.

Keiko: And this privileging of speed, accuracy and using 'the' right method is something that has been critiqued in the mathematics education literature for many years, and which has been seen as a contributing factor to maths phobia. From your interest in the linguistic dimensions of mathematics teaching and learning, how do you see the language, or languages, in the mathematics classroom contributing to, or moderating the kind of inflexible and routine teacher-centred approach to teaching and learning?

Jehad: Thanks for asking this question. I get excited about it. I think that it is so difficult for students to move between the (at least) three forms of language which are in use in Palestine today: the formal Arabic (the written language), the everyday Arabic (the spoken language) and the mathematics itself or what you referred to as a register (noting that this register is multimodal). The textbook presents the written and formal language that students struggle with at the very starting point because it is very different from their daily language which teachers mostly use in explaining the lesson. It is a very complex combination that students have to deal with.

Keiko: It must then be very difficult for students who are learning a new mathematical concept to make meaning in terms of their 'everyday' knowledge if the language to talk about their everyday out-of-school experiences is so different from the written academic language that the textbooks use to introduce and explain the mathematical concepts.

Jehad: Some of those difficulties I explored, not enough though, are that there are mathematical expressions or terms that students struggle with while learning maths. The examples I referred to in some of the publications (Alshwaikh 2011) is that of a 12-year-old student, when I asked him about the relation between the square and the rhombus, and he answered the rhombus 'assists' the square. One way to say the word rhombus in Arabic which is ma'een (معين) which has association with help and assist in daily life.

Keiko: How interesting. So was the student inferring the relationship between the square and the rhombus by using the 'everyday' meaning of the word? Can you then comment on how 'permeable' the walls of the school mathematics classrooms are to the out-of-school lives of the students? Do students' everyday numeracy practices have a place in the school classrooms? Are they acknowledged or valued? Are they utilised as resources for learning the more formal and abstract mathematics in the textbooks? How does/do the allowable language(s) in the mathematics classroom enable discussions about everyday numeracy practices?

Jehad: It is difficult for me to answer these questions. I can tell you how university students struggle with the classical Arabic and how some of them will use the everyday language in their assignments or reports. A second point is the view of maths as a set of symbols, so that students need not use a 'language' to express themselves and/or their solutions. A third point that is relevant here and which is a general comment on Palestinian education is that the classroom is teacher-centred. I mean, the dominant 'voice' in classrooms is the teacher's. In mathematics, students' voices almost vanish because of the dominant view about maths and teaching and learning maths – students need to work on their own and think privately and show their written answers. There is almost no discussion or talk about mathematics in Palestinian schools. This is a claim that needs more exploration and research.

Teachers do not acknowledge the fact that maths is a social practice and this is a key factor which would influence the way teachers engage their students. Here I refer to the kind of final exam that Palestinian students perform at the end of the high school. Many students comment on their grades in maths – that those grades do not reflect their understanding of maths. Rather they reflect their memory of the ideas they learnt. Those students got almost full marks in maths but they described themselves as not strong 'at all' in maths and that they did not understand it. When I asked, 'how did you get such excellent grades?' they answered that the questions in the final exam were very similar to the examples in the textbooks or what teachers would solve but with different numbers – 'all that I need to remember are those examples and then I insert the new numbers'.

Keiko: In the chapter you wrote with Hauke Straehler-Pohl in 2017, you explore the possibilities of political agency in Palestinian school mathematics. Can you talk about what is currently limiting political agency? How would you like to see political agency being more strongly developed in mathematics classrooms?

Jehad: Thank you for this tough question. I do not remember if I had this discussion with Hauke (my co-author) but I have been thinking about it especially within the Palestinian context and after living in South Africa for two years. But I will limit myself to the Palestinian one.

I was convinced that critical education would be *the main* solution for people to engage more (and hopefully change) the political and the social spaces. I have doubts now and I am thinking of what else is needed. Hauke and I in the mentioned chapter (Alshwaikh and Straehler-Pohl 2017) tried to say that it is very complex to describe the Palestinian context, which makes it difficult even to suggest what to do. But let me try and think aloud here.

The critical dimension in education is still needed and necessary but it is not enough. For example, we still need to raise critical questions in maths classrooms as Hauke and I tried to suggest – we need to connect maths to daily life (as Fasheh 1996 has suggested). But it seems that this is not enough in the Palestinian context. So what to do? One suggestion is that we need to understand more about the changes (especially the political and the social ones) that have occurred in this society. Does this just sound like academic rhetoric? Probably yes! I feel hopeless and helpless here. Nevertheless, let me continue. One practical suggestion is what I tried to do – analyse the mathematical discourse within textbooks and challenge the dominant discourse about maths. Another one would be … you know what? I feel that I repeat myself and that I cannot answer this question. I am still trapped in the idea of understanding the changes within the Palestinian society in order to make suggestions. I am so sorry.

Keiko: I don't think anyone would expect a simple answer to this big question! Is there anything else you'd like to add about how your research gives us insights into research in numeracy as social practice?

Jehad: More questions than answers, I am afraid. And this is a reflection on the research as well as your questions in this conversation. How does the use of language and other forms of communication influence the way students learn? I guess that I want to refer to the political situation and agency. How would we convince students and probably stakeholders that language matters, that social practices matter. Again, Hauke and I tried to say that we can teach maths and at the same time be activists and not to choose one of them, as Dowling and Burke (2012) have suggested. I guess that I refer to the broader social/political/cultural/semiotic space in a society. It seems that I suggest that if, at least, we change the way we communicate mathematics (as a social practice) we can do both.

The connections between those ways of communication and their use in education and the social changes in the global sense, while important, still need more exploration and study especially in contexts that are very complicated and have gone through serious changes in a short time such as the Palestinian one.

On the other hand, and to be more optimistic, such research would contribute to the process of understanding the changes in education in such complex contexts. There is a need for quality (mathematics) education in these contexts. I guess that I tried to say that there is a way to look at the nature of maths and mathematical activities that are presented and offered in maths classrooms and textbooks. This may mean that if we know what maths we offer to our students we may be able to offer a different one!

The absence of students' voice in the Palestinian mathematics classroom has been highlighted by Jehad. The silencing of students is effected not only by the nature of the interpersonal and textual functions of the textbooks, but by three other factors. One is the dominant role of the teacher in the classroom leaving little possibility of students to exercise agency. The second is the reliance on the textbooks to guide pedagogy and content of the mathematics classroom. The third is the distinctive registers of the formal Arabic in written texts, the vernacular spoken Arabic and the discipline specific mathematical register. Whereas in languages such as English where even with differences between the spoken and written languages, comprehension of one enables the comprehension of the other, this is not the case in Arabic where there is a very high degree of difference between the spoken and written varieties, including in the syntax. Thus, while the need to learn the language of the discipline (in this case mathematics) is common to the school students' experiences in both the Arabic-speaking and English-speaking contexts, the students that Jehad describes in his contexts are also learning to negotiate the formal written variety of Arabic. In the English speaking context, the written and spoken English can both mediate the learning of the technical mathematical language. However, in the Arabic-speaking context, the students have only the spoken language to mediate the learning of both the academic and the technical mathematical languages. The situation is exacerbated, however, by the limited opportunities for students to use 'classroom talk' to even undertake these tasks.

Conclusion

The meaning-making practices in the Palestinian mathematics classrooms that Jehad has painted is one that has been powerfully shaped by the textbooks that have travelled from afar, bringing with them a particular, but familiar, ideology about what constitutes legitimate mathematical knowledge and ways of knowing it. While mathematics as a semiotic system is multimodal, only particular types of texts are currently allowed to enter the classroom. Published textbooks dominate, limiting the meaning-making potentials to those afforded by the difficult written variety of Arabic and the technical language of mathematics, and teacher talk that replicates what is in the textbooks. The exclusion of student talk using their everyday spoken Arabic means that there is little opportunity for students' everyday experiences – the meaning-making resources that the students have – to enter the classroom. A way of understanding what is happening in the Palestinian mathematics classroom is to say that aspects of the global represented by the textbooks, and aspects of the local represented by the Arabic diglossia, work together to preserve the hegemonic nature of the autonomous model of mathematics.

While the situation does not exude optimism, the analysis such as what Jehad has been undertaking to examine the affordances (or otherwise) of different modes of communication in the classroom enables us to identify the potential for change. For example, introducing more communicative pedagogies that encourage student talk may begin to bring the students' own meaning-making resources to enter the classroom, and unleash some of the unrealised agency of students. This may be what is needed to 'rehabilitate' the hegemonic discourse about mathematics that has been the feature of Palestinian classrooms. This is no simple project, but points to a potential 'limit of the global' that can be examined further.

Notes

1   The Bourbaki group was a group of young, mainly French, mathematicians who published a number of mathematics texts starting in the mid-1930s, with a view to modernising the treatment of mathematical ideas in textbooks.
2   Road number 443 is an apartheid road. Jews: Palestinians refer to the Israeli soldiers and settlers as the Jews. Abu Mazen: is the Palestinian president whose term expired in 2009. Fateh: is the main political party who controls the Palestinian Authority in the West Bank.

References

Alshwaikh, J. (2011) "Geometrical diagrams as representation and communication: A functional analytic framework". PhD thesis Institute of Education, University of London, London.
Alshwaikh, J. and Straehler-Pohl, H. (2017) "Interrupting passivity: Attempts to interrogate political agency in Palestinian school mathematics". In Straehler-Pohl, H.,

Bohlmann, N. and Pais, A. (eds), *The Disorder of Mathematics Education: Challenging the Socio-Political Dimensions of Research* New York; Springer.

Baker, D. (1998) "Numeracy as social practice" *Literacy and Numeracy Studies*, 8(1): 37–50.

Barwell, R., Clarkson, P., Halai, A., Kazima, M., Moschkovich, J., Planas, N., Setati-Phakeng, M., Valero, P. and Villavicencio Ubillús, M. (eds), (2016) *Mathematics Education and Language Diversity* Cham: Springer International Publishing.

Brandt, D. and Clinton, K. (2002) "'Limits of the local': expanding perspectives on literacy as social practice" *Journal of Literacy Research*, 34(3) 337–356.

Dowling, P. and Burke, J. (2012) "Shall we do politics or learn some maths today? Representing and interrogating social inequality". In Forgasz, H. and Rivera, F. (eds), *Towards Equity in Mathematics Education: Gender, Culture, and Diversity* Heidelberg; Springer.

Fasheh, M. (1996) "The main challenge: Ending the occupation of our minds. The main means: Building learning environments and re-contextual knowledge". In *Political Dimensions of Mathematics Education III*. Bergen: Caspar Forlag.

Halai, A. and Clarkson, P. (2016) *Teaching and Learning Mathematics in Multilingual Classrooms* Rotterdam: SensePublishers.

Halliday, M. A. K. (1978) *Language as Social Semiotic: The Social Interpretation of Language and Meaning* London: Edward Arnold.

Halliday, M. A. K. (1993) "Towards a language-based theory of learning" *Linguistics and Education*, 5(2): 93–116.

Iedema, R. (2003) "Multimodality, resemiotization: extending the analysis of discourse as multi-semiotic practice" *Visual Communication*, 27(1): 29–57.

Kress, G. (2010) *Multimodality: A Social Semiotic Approach to Contemporary Communication* London: Routledge.

Kress, G. and Van Leeuwen, T. (1996) *Reading Images: The Grammar of Visual Design* London: Routledge.

Latour, B. (1987) *Science in Action: How to Follow Scientists and Engineers through Society* Cambridge, MA: Harvard University Press.

Latour, B. (1990) "Drawing things together". In Lynch, M. and Woolgar, S (eds), *Representation in Scientific Practice,* Cambridge MA: MIT Press.

Lerman, S. (2000) "The social turn in mathematics education research". In Boaler, J (ed.), *Multiple perspectives on Mathematics Teaching and Learning* Westport, CT: Ablex.

Morgan, C. (2006) "What does semiotics have to offer mathematics education research?" *Education Studies in Mathematics* 61: 219–245.

9

TEACHING AND LEARNING OF NUMERACY IN NEPALESE PRIMARY SCHOOLS

Mariko Shiohata[1]

Introduction

Over the past three to four decades, many who work to promote education in low-income countries have been concerned with the quality of teaching and learning. In recent years, however, the learning outcomes achieved in many such countries have been more frequently and prominently highlighted as worryingly poor. For example, in 2014, UNESCO reported that about 250 million children globally did not have basic skills in literacy and numeracy, and, disturbingly, 130 million of them were unable to read or calculate even though they were attending school (UNESCO 2014).

The Millennium Development Goals and the Education for All agenda, both started in 2000, encouraged governments to abolish fees and move towards universal primary education. As a consequence, many schools ended up having far more children than they could cope with, leading to unsatisfactory levels of learning. Thus, the expansion of access can, in some instances, bring about negative quality effects (Somerset 2011).

This so-called 'learning crisis' of recent years was repeatedly featured in the preparatory debates in the run-up to the adoption of the Sustainable Development Goals at the United Nations General Assembly in 2015. In this new global agenda, the international community agreed to strive towards a broader set of goals for education, encompassing quality as well as access:

> By 2030, ensure that all girls and boys complete free, equitable and quality primary and secondary education leading to relevant and effective learning outcomes.
>
> *(United Nations 2015)*

Agreeing to recognise the goal of ensuring learning for all children is one thing, but delivering it is another matter. Some research institutions, development agencies and international non-governmental organisations (NGOs) have turned to the international assessment surveys such as Programme for International Student Assessment (PISA) and Trends in International Mathematics and Science Study (TIMSS) as instruments with which to make judgments as to the effectiveness of young people's learning, and by extension, the quality of education systems (Barrett 2016). One such example was the Learning Metrics Task Force (LMTF), an inter-agency group for the promotion of measurement of learning led by the UNESCO Institute for Statistics and the Brookings Institution. It advocated the establishment of a set of 'skills and competencies that are important … and a small set of indicators that are feasible and desirable to track at the global level' (UNESCO, UIS and Centre for Universal Education 2013, 10).

However, the simple emphasis on tracking children's learning levels runs the risk of reinforcing a narrow focus on testing, aimed purely at the assessment of learning, not on the *improvement* of learning (Archer 2014). The danger in placing a heavy emphasis on learning outcome goals is that it may lead to reliance on high-stakes summative testing, which can be detrimental to the meaning and relevance of education (Barrett 2011). To build the link between assessment and improved learning, the processes of pedagogy which take place in the classroom need careful and critical attention (Alexander 2015).

Although the processes of teaching and learning have a deep significance for children's learning, they are hard to capture. Most aspects of the educational process can only be judged through observation against qualitative indicators (Alexander 2008; O'Sullivan 2006). It is not realistic to attempt to incorporate these processes into macro-level evaluation of quality.

The challenge now facing education policy-makers and practitioners is to explore ways in which knowledge and understanding of learning outcomes can be used for diagnostic purposes to improve teaching and learning. David Archer of ActionAid advocates 'a massive complementary investment in formative assessment by teachers themselves' (Archer 2014). Teachers need to be able to identify the progress their students are making, and the learning difficulties they are encountering, and to adjust and modify their teaching as needed to meet these difficulties. In many low-income countries, however, the hasty expansion of enrolments has led to the deployment of large numbers of non-professional teachers who have not been trained even at a basic level (Lewin 2007). Teachers' capacity to provide quality instruction in the classroom is the key to linking assessment to learning improvement, but the necessary enabling environment is absent in many countries.

This chapter discusses some results from a basic education project in Nepal which attempted to enhance teachers' skills in understanding their students' learning of basic mathematics. The project started in 2012 and ended in 2015. It builds on a previous initiative, the Continuous Assessment System (CAS), introduced in the 1990s to complement the Liberal Promotions Policy (LPP) which aimed at reducing drop out and repetition during the five-year primary education cycle (World Education Forum 2000).

When CAS was introduced, a form for recording students' progress was developed as a key component, and circulated to the schools. CAS form filling has become a routine for many teachers, but few of them utilise the information to identify gaps in learning or to modify their teaching accordingly. The basic education project attempted to remedy this deficiency. Our aim was to move teachers away from a preoccupation with record-keeping, and to draw their attention towards the importance of oral questioning and diagnostic testing, as sources of information about their students' learning difficulties, and as a basis for giving constructive feedback. As the project developed, we started to realise that teachers' beliefs concerning children's learning are a critical factor in determining the quality of the instruction they provide.

Perspectives on classroom instruction

Within the debate on what constitutes education of good quality and how to make it materialise for children, there is consensus that children must acquire the basic and minimum competencies in literacy and numeracy. These skills form the foundation on which later progress in learning is based, and those children who fail to achieve sufficient progress are likely to drop out or encounter difficulties in understanding contents of the other subjects (Alexander 2015; Dubeck and Gove 2015; Somerset 2016).

For some time, there has been a consensus that formative practices enhance students' learning. But while there is a large body of literature in the global North, accounts of formative assessment practices in low-income countries are relatively limited, except for cases from Ghana (Akyeampong et al. 2006), Malawi (Hares 2013), South Africa (Lumadi 2011; Pryor and Kubisi 2002), and Zambia (Kapambwe 2010).

In Nepal, the words 'formative assessment' and 'continuous assessment' were frequently used among education department officials, aid agency staff, and other education workers, but confusion surrounding their meaning and purpose was persistent.

It is evident that in developing the CAS Implementation Book, the staff of the Curriculum Development Centre (CDC) intended it as a teachers' guide in the use of students' assessment data for formative purposes (CDC 2011, 5):

> CAS is a system that goes hand in hand with the teaching and learning process. For effective teaching and learning, we can't separate assessment process from teaching and learning process.

Assessment, it is suggested, should be an integral component of the teaching and learning process: it should accompany pedagogy rather than follow it. This view resonates with the definition of formative assessment given by Alexander (2015, 254):

the day-to-day, minute-by-minute observations and interactions through which good teachers constantly monitor children's learning and progress, affording the feedback which will build on their understanding and probe and remedy their misunderstanding.

However, observations in the field in Nepal showed that formative elements have been lost and the focus has moved to summative record-keeping, leaving many teachers puzzled as to the relationship between records of student progress and planning for future lessons.

The CAS form devised by CDC specified that the teacher should evaluate the students on the basis of various criteria such as participation in classwork, project work, creative activities, and behavioural change. Each student received a number of tick marks – between one and three – on each criterion. However, these criteria were not clearly defined, and guidance as to how the information could be gathered, and how it could be used for strengthening future teaching, was not provided. Apparently it was assumed that the teacher could devise these crucial steps for him/herself.

This shift of focus in Nepal from formative to summative assessment, and towards a heavy emphasis on record-keeping, is strikingly similar to experience in Ghana (Akyeampong et al. 2006) and Malawi (Hares 2013). A radical reformulation of the roles and responsibilities of teachers, coupled with systematic teacher in-service training was needed for the effective implementation of CAS. However, these essential steps were not taken. In the field a 'CAS implementing school' was taken to mean a school where teachers use the CAS form or practise some kind of record-keeping. It was impossible to gauge whether, and to what extent, continuous assessment was being effectively practised without visiting the school and observing lessons.

In teacher preparation programmes in many low-income countries, very little time is given to learning how to teach. Rather, by far the largest amount of time is allocated to developing trainee teachers' subject content knowledge (Lewin and Stuart 2003), presumably on the assumption that once the new teacher has acquired this subject knowledge, pedagogical skills will follow more or less automatically. This is of course a fallacy: the gap between the competencies teacher trainees need to become effective pedagogical professionals and the preparation they receive during most initial training programmes is a wide one (Akyeampong et al. 2013). In Nepal, the rapid expansion of basic education has contributed to the worsening of the quality problem. The number of permanent teachers has been frozen, but an increasing number of teachers has been recruited on a contract basis, often without receiving any systematic training.

The most common approach to teaching numeracy observed in the classroom was simply to impart content knowledge. Lessons typically began with a demonstration of an algorithm on the chalkboard, followed by written exercises based on it for the students to carry out individually. The abler students came forward as they completed the task and the teacher marked their work, while the other students continued struggling. In many cases, the teacher ignored these students, and moved on to the next task before they had the chance to complete the first task. Conspicuously missing in these lessons was any attention on the part of the teacher to the learning process, and more specifically to differences in levels of comprehension among his/her students.

Method

In 2012, Nepal's Department of Education (DOE) of the Ministry of Education launched a basic education development project in collaboration with Save the Children, focusing on five of the country's 75 districts. A recently published mid-term evaluation report of the ongoing School Sector Reform Program (SSRP)[2] had highlighted issues concerning the implementation of Continuous Assessment System (CAS). Both parties agreed to make efforts to strengthen CAS, and more generally, to improve teaching and learning in the project areas.

A series of training programmes was designed for delivery to resource persons (RPs)[3] and teachers in the project schools. The implementation team consisted of DOE and Save the Children staff. The training workshops were organised in two phases in each project district, over a period of one and a half years. In the first phase, a strong focus was put on formative assessment: activities undertaken by teachers to gather information about their students' learning, and use it to modify their teaching. For this purpose, much attention was paid to the development of teachers' oral questioning skills, as a key tool for strengthening pedagogy. One trainer acted as a teacher, and the participants as students, to demonstrate effective questioning techniques. A video clip showing a Nepalese teacher asking low-order and high-order questions of her students was screened. (All training sessions were conducted in Nepali, the official language of the country, with the exception of a couple of presentations about formative assessment given by the author. These were conducted in English and translated into Nepali.)

In each district, the first-phase programme was typically structured along the following lines:

- *Day 1:* Discussion of the meaning of CAS, formative and summative assessment, purposeful use of question-asking in the classroom.
- *Day 2:* Teaching technique with formative element, demonstration teaching preparation, introduction to a number skills diagnostic test.
- *Day 3:* Visit to schools for demonstration teaching by participants, administration of the diagnostic test.
- *Day 4:* Reflection on the demonstration teaching, preliminary analysis of results of the diagnostic test, discussion of meaning of education and learning, development of action plans for changes to pedagogy in the participating schools.

For training purposes, several tools were devised by the trainer team. One of these tools was a basic number skills diagnostic test, which was introduced and administered in two out of 40 project schools in each participating district. Following the lead of Somerset's (2003) work in the Philippines, the test was based on Nepal's current math curriculum for the first three primary grades, and designed to identify the particular concepts which students find difficult to comprehend, and to enhance teachers' awareness of their students' systematic math errors.

The test was made up of four main sections, as follows:

1   *Counting:* Six questions tested the simplest number skills. Students were shown groups of apples, numbering between 1 and 9, and asked to count how many apples there were in each group.

2   *Number values:* Students were asked to arrange groups of three numbers in order, from the smallest to the largest. In the first two questions the numbers all consisted of single digits, but the later questions included three- and four-digit numbers.

3   *Mechanical arithmetic:* A third group of questions tested the students' ability to carry out the four basic number operations – addition, subtraction, multiplication and division. The simplest questions involved one- and two-digit numbers only; the more complex questions, three- and four-digit numbers, together with carrying and borrowing.

4   *Applied number problems:* The final group of questions presented students with practical number problems, of the kind frequently met in Nepalese everyday life. In number problem questions, students are not told which operations to carry out, as they are with mechanical arithmetic questions. Instead they must decide for themselves, from the information given, which operations are needed. The training team constructed appropriate questions drawing on the knowledge of the Nepalese contexts. Some examples are:

    o   Ramesh has 25 ducks. He sells 6 of them. How many ducks does Ramesh have now?

    o   There are 3 football teams in Shree Narayan School. Each team has 11 players. How many football players are there in the 3 teams?

    o   Sita goes to the shop and buys 1 kg of rice for R 65. She pays with a R 100 note. How much change should she get?

When the teachers in the participating schools were shown these questions, they responded that their Grade 5 students should be able to answer correctly.

The second phase, which was held about ten months after the first workshops, focused on a review of the progress achieved by the participants in implementing the techniques demonstrated during the first-phase programme. Aiming at enhancing the participants' understanding of continuous assessment, the problems the participants faced in implementing continuous assessment in the classroom were discussed.

The second-phase programme was structured as follows:

- *Day 1:* Review of progress of CAS implementation in schools, review of two types of assessment (formative and summative), written question construction.
- *Day 2:* Briefing about CAS by Curriculum Development Centre (CDC), analysis of records of effective and not-so-effective lessons, preparation for demonstration teaching.
- *Day 3:* Visit to schools for demonstration teaching by participants.
- *Day 4:* Reflection on the demonstration teaching, discussion of students' systematic errors in mathematics from the previous year's test results, development of further action plans.

After the diagnostic test administration during the first phase, the trainer team wrote a report, analysing the different kinds of errors the students had made in answering each type of question. The test results showed that in some schools as many as two-thirds of the participating students responded that 168 was larger than 201. This indicates that they compared the two numbers and saw that 168 had larger digits than 201, indicating their misunderstanding of number place values. Another error frequently made by the students was in carrying and borrowing. Two examples will be shown later in Figure 9.1.

The report was made available to the participating schools during the second phase as guidance feedback. At each school, Save the Children's project staff explained the contents of the report to the head teacher and teachers, and encouraged them to think why their students made these mistakes and how they might modify their teaching accordingly. In most cases, the teachers accepted the feedback positively, sometimes showing surprise at their students' mistakes or noting that they had never received such feedback before. This was a major shift for teachers who had, for the most part, been used to simply checking the students' answers as right or wrong.

Examples of systematic errors were discussed during the second-phase workshop for wider sharing with all the other participants, and the reasons why students make such mistakes analysed. Some of the teachers from the participating schools talked about their experiences, giving accounts of formative and diagnostic actions they had started to take in their teaching.

Findings

The findings discussed below derive from the following sources:

- Focus group discussions on continuous assessment with teachers participating in training during the first phase of the workshop.
- Lesson observations at the schools participating in the math test during and after the training workshops.

- Open-ended interviews with the head teachers and teachers of the schools where the math test was administered.

The open-ended interviews were conducted to investigate teachers' beliefs and knowledge about assessment and to explore their perceptions of practice. An interpretive, case-study approach was employed. During these interviews, the math test results were used to elicit teachers' beliefs about teaching and learning, and their perceptions about the causes of student failure.

The CAS forms

The evidence showed that the aim of getting teachers engaged with continuous assessment was unlikely to be achieved simply through the circulation of CAS forms. As part of the original CAS implementation, schools were encouraged to abolish annual exams and to promote all students automatically to the next grade. However, little explanation was given as to how use of the CAS performance records could replace the examinations. More importantly, there was no clear guidance to teachers that students needed to reach a certain level of learning at the end of the year before being promoted. Head teachers and teachers tended to confuse CAS forms with exams, whereas the district officials blamed the government for not establishing a proper mechanism to implement continuous assessment.

The first-phase workshop started by putting the question, 'what is CAS?', to the participants before any explanation about it was provided. The trainer team made the judgement that it would be useful for participating teachers to discuss conflicting perceptions of CAS freely to generate an understanding of the purposes and meanings of assessment.

The response of about a quarter of the 40 participating teachers in District A was that CAS is for measuring completed learning. One male head teacher said: 'CAS is about giving tick marks after measuring the students' achievement' (17 June 2013). Similarly, a male teacher in District B responded: 'After a lesson, students are asked questions and based on their answers, they are given tick marks' (12 August 2013). It was clear that these teachers made no distinction between CAS and periodic tests; in their perception, entering ticks on the CAS forms was simply another form of summative assessment. Some teachers conducted tests, oral or written, before deciding the number of tick marks to give to the students in completing the form. They then accumulated the sum of the tick marks to decide the final grading at the end of the school year. Some teachers told the author that they were not sure as to whether and in what way the total sums of tick marks for individual students could be made compatible with exam results.

A few teachers, however, had understood the formative purposes of CAS. These teachers expressed the view that CAS is concerned more with teaching and learning, rather than with summing up students' achievement: 'CAS can be useful to find out the status of the students and to improve teaching itself'; '[CAS is] assessment during the teaching learning process' (District B, 12 August 2013). Nevertheless the majority of the teachers' views of CAS were confused, indicating poor understanding of the meaning and purpose of continuous and formative assessment: 'CAS helps students achieve learning outcomes and become more disciplined' and 'Promoting students by assessing their attendance and involvement in extracurricular activities' (District A, 17 June 2013); 'After [conducting] continuous assessment for a month, we should [be able to] identify who stand first, second and third positions in the entire class' (District C, 30 August 2013).

Most teachers in the training workshops felt that their students' current levels of achievement were lower than they should be, and that therefore some measures were needed. However, they usually blamed others for this state of affairs. Some criticised parents for not showing a sufficient understanding of the education of their children, whereas others pointed to district education office's unsatisfactory backing and support. Few of them problematised classroom practices or made a connection between teaching process in the classroom and learning improvement. Nevertheless, many teachers shared an assumption that once 'modern' teaching methods are adopted, learning will be strengthened. They often contrasted 'traditional' approaches to teaching with 'modern' approaches, frequently associating the latter with student-centred pedagogy. As a male teacher said:

> Teachers don't understand its [CAS's] importance because of lack of training, resources and time. Many have just traditional mentality. We can't use student-centred pedagogy. Teachers and students are used to doing written exams only.
>
> *(District B, August 2013)*

Oral question-asking

There is growing evidence from around the world that teachers' use of interactive communication strategies enhances classroom pedagogic practices and has a positive impact on student learning outcomes (Westbrook et al. 2013). These practices, according to Alexander (2001), comprise: 1) teachers' spoken discourse; 2) visual representation; 3) the act of setting or providing tasks for learners; and 4) a variety of social interactions.

The training workshop team paid particular attention to teachers' spoken discourse, including instruction, explanation, questioning, responding, and elaboration. In the majority of the classes in the project schools observed by the trainers during the preparation period, many teachers just read from the textbook and did not ask questions of the students, nor encourage them to be engaged with the lesson. Even when they did ask questions, teachers generally addressed them to the whole class, limited them to low-demand 'what' questions, and accepted a choral response. The purposeful use of oral question-asking techniques was rare. According to one study conducted in the early 2000s in Nepal, in only 10 per cent of the observed lessons in 19 schools did the teacher use the strategy in which she questioned the class, paused for students to think, and then indicated one student to respond (Research Centre for Educational Innovation and Development 2004). This question presentation strategy is argued to be more effective than other strategies as it encourages all students in the classroom to think about the problem and work out answers for themselves. In the training workshop, this question presentation strategy was called APPLE: *Ask the question; Pause*, letting the learners to think about what you are asking; *Pick* on a learner by name to answer the question; *Listen* to the answer, make eye contact with the learner, provide effective words when the answer is provided; and *Expound and explain the learner's answer.*

In introducing this strategy to the participants, the trainers stressed that the teacher employing it should not always choose the first student who raises his or her hand, but rather wait until the majority are ready to respond. The teachers were advised that occasionally the teacher may also choose a student who has not raised his hand to encourage participation. They were also told that it is important to generate a dialogue based on the learner's response; if the student's response is incorrect, the teacher should redirect the question back to the other students, by saying for example, 'That's an interesting response, but not the one I was looking for, can anyone else provide a different answer?'

In both phases of the workshop, the trainer team emphasised the effectiveness of this strategy and encouraged the teachers to use it in their demonstration teaching. The team's expectation was that the teachers would adopt it readily, because no special preparation was required. In the event, however, this proved not to be the case. Most teachers were so used to asking questions of the whole class and to accepting choral responses, that few of them were able to adopt the new strategy effectively. Similarly students were so used to responding to questions immediately, individually or more often in chorus, that teachers found it difficult to get them to wait until asked to respond.

Between the first- and second-phase workshops, the author visited a number of schools in search of evidence of effective use of the new approach to questioning. Many teachers claimed that they had started asking questions in this advocated way, but lesson observations showed that in most cases, this meant that the teachers had started directing questions to individual students. Rather than addressing the question initially to the whole class to encourage all the students to think, many teachers first nominated the student to respond before asking the question. Clearly, the rationale for the changed approach had not been understood.

In general, there seemed to be an absence among the teachers of willingness to let the students think. Bruner's (1996) four models of learners' minds are relevant here:

1   seeing children as imitative learners;
2   seeing children as learning from didactic exposure;
3   seeing children as thinkers;
4   seeing children as knowledgeable.

Although the participants had been exposed to the notion of child-friendly schools through training by external aid agencies, their dominant beliefs about children's learning were clearly more compatible with the first two models, rather than with the latter two. The trainer team implicitly expected that the teachers would start espousing the third and fourth models and therefore adopt the more effective questioning strategy. But this is a fundamental transformation because the teachers are expected to shift their existing notion that children are empty vessels to receive knowledge passively to the new notion that children need to think actively so that they understand; it is therefore not one which is likely to be brought about easily or quickly, through a couple of training workshops.

Nevertheless, some teachers did use the questioning techniques promoted in the training to greater or lesser degree. One lesson observation taken from the author's field notes is given in Box 9.1. At this school, the math diagnostic test was administered to the 47 Grade 5 students. Immediately after the test was completed, the math teacher was asked to review some difficult questions with his students. He himself had not participated in the training workshop, but nevertheless practised purposeful question-asking.

Math diagnostic test and feedback

The importance of giving feedback to students to enhance their learning is increasingly recognised and there is a large body of literature on this (e.g. Hattie and Timperley 2007). However as we have seen most training participants' day-to-day feedback to their students was limited to checking simply their answers as right or wrong, almost never looking into why particular mistakes were made. The math diagnostic test was introduced with the aim of enhancing teachers' awareness of students' systematic errors, and subsequently strengthening their skills in giving feedback. When examples of these systematic errors were shown to the teachers during the first workshop, they often came as a surprise and shock to them. Some participants said that the students did not think enough before answering; others that many students have not understood the concept of carrying and borrowing.

As shown in Figure 9.1, the diagnostic questions used the Nepali numeric symbols, in accordance with the national textbooks. It was noticeable, however, that in some cases students were not consistent in choosing a script for writing their answers. Some wrote entirely in Nepali script; others wrote entirely in Arabic script, while many mixed the two scripts. Furthermore, numbers in Nepali script were often written in ways which could be confused with Arabic

BOX 9.1   Math teaching at a school in District E

After conducting the basic number skills test with 47 Grade 5 students, I picked out several questions which are usually difficult for many students. These questions were:

$$47 \qquad\qquad 675 \qquad\qquad 812 \div 4 =$$
$$\underline{\times\, 8} \qquad\qquad \underline{-\,493}$$

I copied these questions on the chalkboard and asked the teacher, who has been teaching math in this school for the last four years, to explain to the students. He solved the first two questions with the students, addressing the whole class, and carrying out each operation with them. For the first, he started by asking what he should write below '7' and '8', pointing to the two digits. The students answered '6' in unison. He then asked whether there is a digit to be carried, and the students responded '5!', again in unison. They reached the answer 376. He solved the second question with the students in a similar manner.

For the third and last question, he pointed to one female student, asking her to solve it in front of everyone. She came to the chalkboard and started the division very cautiously. She re-arranged the numbers by putting '4' on the left side of '812' and drew lines for division. While she was doing this, everyone was watching her work quietly. The teacher did not rush her at all. After the student calculated the correct answer (203), he asked another student (a boy) to check the answer. This checking was not simply a matter of 'yes' or 'no'; he was asked to calculate at his desk to see whether he reached the same answer. The student said he did. The teacher nodded.

Realising that this problem was a little tricky, the teacher then gave another, similar question (not from the diagnostic test), which was $402 \div 2$. He pointed again to one student and asked him to solve it on the blackboard. He reached the correct answer.

From the above transactions, it was clear that most of the students were following what was happening.

Then I gave the class two further questions from the diagnostic test, which were:

$$39 + 282 = \qquad\qquad 4138 - 753 =$$

For the first question, the teacher arranged the two numbers (39 and 282) vertically, asking his students which digit should come at each place value. They reached the correct answer jointly.

Then, he pointed to a female student who was sitting at the back of the room. She came to tackle the second question on the blackboard. However, she looked at the problem for a while, doing nothing. While this student was struggling, the teacher quietly brought another student, and asked her to try. This student re-arranged the numbers vertically, with all the digits at the correct places. The teacher did not say anything, but the first student quietly saw her peer solve the problem. Although this female student was not asked to try again, there was absolutely no humiliation for her.

When the second girl solved the question correctly, the teacher pointed to another female student and asked whether the answer was correct. She checked by calculating on her notebook and said, '3385'. The teacher again asked another student, a male, whether the answer was correct. This student said, 'I'll check it by writing on the blackboard', and copied the problem in Nepali numerical symbols. He carefully carried out the operations and reached the same answer.

Throughout this teaching which took only about 15 minutes, the teacher was calm, cool, and even quiet. There was absolutely no shouting or raising of voice by the teacher or any student. It was evident the teacher uses various question-asking techniques effectively.

Initially he accepted choral responses but later began addressing the question to the whole class, then nominating individual students to respond. At one point he skilfully included a weaker student. Even though this student could not answer correctly on this occasion, it is likely that she will become aware that she needs to think for herself, because she was not punished or insulted, but encouraged to take part in the lesson. The quiet environment allowed the students to think carefully.

(from field notes, 29 September 2013)

numbers. In particular, Nepali 4 often looked like Arabic 8; Nepali 7 like Arabic 6; and Nepali 8 like Arabic 2. It is clear that the acquisition of basic numeracy would be potentially enhanced if a standard set of number symbols could be agreed, and applied universally in all schools by all teachers.

Some teachers asserted that the errors shown to them did not fairly reflect the tested students' abilities. This suggests that they regarded any kind of paper test as summative assessment, even though the facilitator team repeatedly explained that the purpose of the test we had used was formative and diagnostic.

In recent years, use of locally available or produced learning materials has been enthusiastically promoted both by the government and aid agencies. In maths lessons in schools, the author observed that some teachers were using small stones or pebbles to teach counting of small numbers, as well as addition and subtraction. Such materials can be highly effective to enable children to

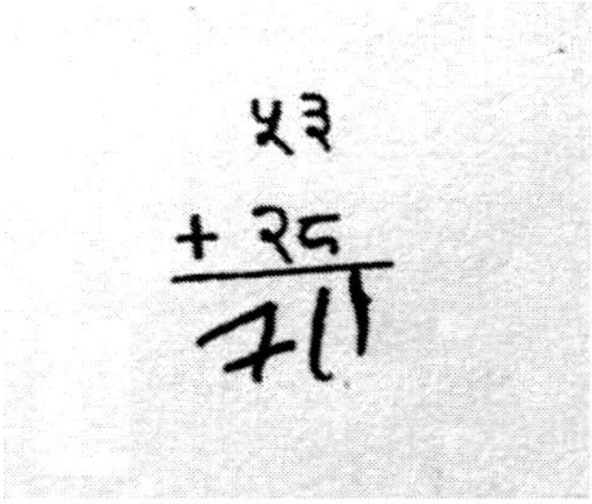

In the questions involving addition and multiplication, carrying errors were very common. In the first question (53 + 28), for example, the student has started by adding the digits 3 and 8, giving the sum of 11. However, instead of carrying the tens digit to the second column, he has written both digits into his answer line. He then adds 5 to 2, giving 711, instead of 81, as his final answer.

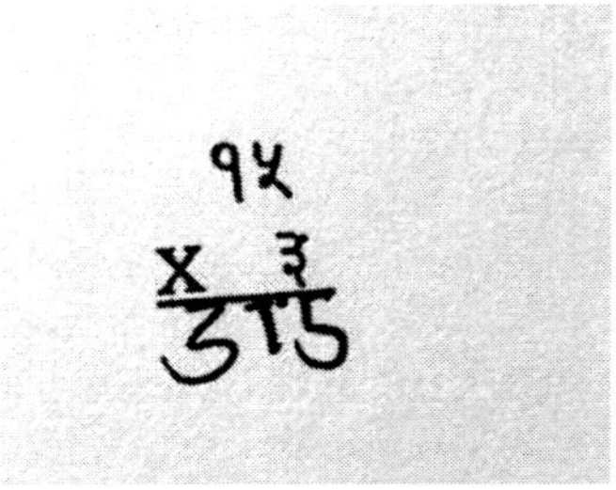

In the second question (15 x 3), the student multiplied 5 by 3, giving the sum of 15, and instead of carrying the tens digit to the second column, he wrote both digits into the answer line. He then multiplied 1 by 3, giving 315, as his final answer.

FIGURE 9.1   Examples of student systematic errors

understand the basic number values, but are not suitable for learning large numbers. However, it was clear from the test protocol that some students were using counting rather than calculation to arrive at their answers. In Figure 9.2, the student drew bars to show his counting process.

The schools which participated in the test subsequently received a feedback report. All of them responded to the feedback positively. One female teacher in District B said:

'We cannot necessarily become aware of our problems [in teaching]. We need guidance from outside too' (January 2014). Another female head teacher in District C told us that the diagnostic test had changed her ways of understanding children's learning:

> After receiving the math diagnostic feedback report, we realised that this is a real problem. We realised that we were not teaching in the ways students could understand. To address the problem, we started doing mini-tests regularly, not only in math but in all subjects. Previously, I and other teachers were aware of students' mistakes. We talked about those mistakes in the staff room, but didn't think that we needed to take action. We were thinking students make mistakes because they don't work hard enough or because they have conceptual problems. After receiving CAS training and subsequent feedback on math systematic errors, we realised that feedback is necessary. We can't always do individual feedback, but when some mistakes are shown using the chalkboard and explanations are given, children respond and make corrections themselves. (24 May 2014)

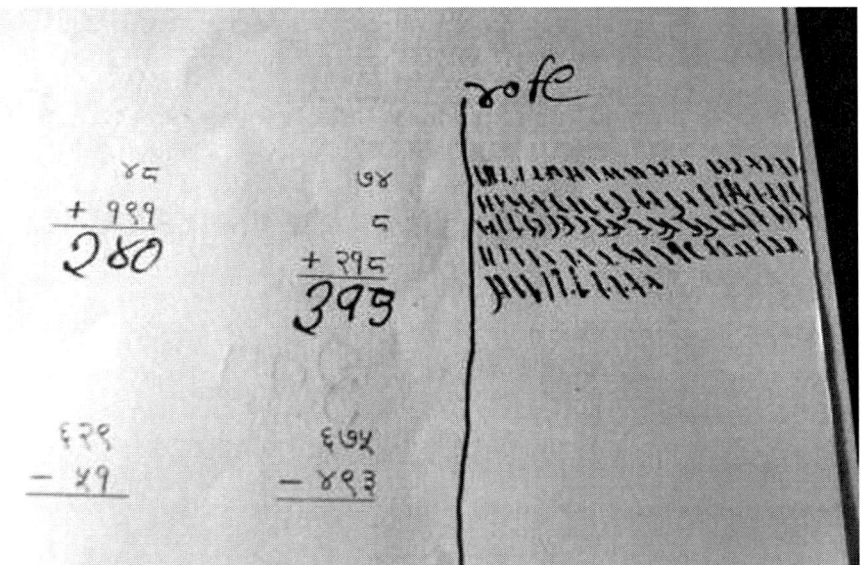

FIGURE 9.2   A student's counting process

This case illustrates that, for many teachers, once they find a particular strategy meaningful, they are likely to adopt it. They start to feel that their students are responding positively to feedback, and that this has resulted in more meaningful teaching and learning. In a school in District A, students' test responses showed a high level of understanding of basic math concepts. In his interview with us, their teacher noted:

> I don't do anything special. I try to ensure that the students understand number place values and memorise multiplication tables. During lessons, I involve students in solving problems by asking them to come to the chalkboard. I have education background, and understand that children have their own individual learning needs, therefore I need to deal with them individually. It's important to understand students. For that purpose, it's important to ask questions when I introduce new concepts to them. (August 2014)

In other schools, the teachers had also responded to the initial feedback positively, but nevertheless had not begun changing their teaching practice accordingly. Not surprisingly, student learning had not been enhanced in these schools. Asked as to whether he had done anything differently after receiving the feedback report, one such teacher said:

> I tried. I mixed weak students with able ones so that able ones can support the weak ones … I don't know … I may not possess a technique, but

my students can answer correctly during the class, but they can't do it in exams. (12 August 2014)

This teacher is clearly struggling to understand why his students' learning is not satisfactory. Although he says that his students can give correct answers during lessons, observation of his pedagogy showed that his monitoring of student learning was incomplete and ineffective. He checked only a small proportion of students' workbooks, and never used incorrect answers formatively.

Conclusion

This chapter has pointed to a disjunction between the policy vision for continuous assessment and the realities which constrain its implementation at the school level. The introduction of Continuous Assessment System in Nepal was largely limited to the circulation of the paper recording form. Not surprisingly, this step by itself did not lead to real change in classroom practice nor to learning improvement. In the absence of guidance as to the meaning and purpose of formative assessment, most teachers took the completion of the CAS forms simply as a substitute for setting and marking examinations.

The implementation of the new policy was further constrained by a tendency, on the part of many teachers, to ascribe responsibility for learning problems to factors external to themselves – to the lack of resources, the laziness or lack of intelligence of the students, or the indifference of their parents. Accepting that the quality of their pedagogy also contributes to student failure is a step that many teachers find difficult to take. However, the chapter also highlighted a small number of cases in which teachers adopted new approaches after realising that the ways in which they had been teaching were not effective.

Numerous attempts have been made, and substantial resources allocated, to introduce what is broadly described as 'learner-centred education (LCE)' into low-income countries in the past few decades. However, the history of the implementation of LCE is 'riddled with stories of failures grand and small' (Schweisfurth 2011, 425). Insufficient consideration for cultural and historical diversities and local contexts undoubtedly has played some part in these unsuccessful attempts. If particular LCE strategies are at odds with teachers' beliefs about learning, attempts to introduce constructivist ideas are likely to have little effect on their practice. As we have seen, the current perception of children as 'empty vessels' remains the dominant pedagogical stance of teachers in our project areas.

However, this does not necessarily mean that these are authentically traditional Nepalese values and therefore immutable. There are already teachers who practise purposeful oral question asking and who provide effective feedback, regardless of their training experience. If, through sensitive handling, constructivist pedagogies are adapted to the local context, they are likely to take root, albeit slowly and gradually. Given training programmes that intertwine the theoretical and practical elements, teachers might start to integrate the knowledge they acquire from training into their experiential knowledge of specific classrooms and students (Lewin and Stuart 2003).

The approaches espoused in this chapter resonate with what O'Sullivan (2004) calls a shift from learner-centred to learning-centred pedagogy, in which knowledge and skills are constructed through focused dialogue between teacher and students (Alexander 2001).

Teachers are the key to the success of any learning improvement initiative. Yet they are often problematised as a barrier to the enhancement of education quality, rather than as an indispensable resource. If future attempts to reform teacher education are to bear fruit, more evidence-based policy, backed up by rigorous and context-based research, will be needed. Generating a better understanding of the actual ways in which teachers acquire and use their professional knowledge will be a key component of such a strategy.

Notes

1 Views expressed in this chapter are the author's and do not necessarily reflect those of Save the Children.
2 The SSRP was a six-year education plan started in 2009, aiming at comprehensive education reform. A mid-term survey was conducted by a group of consultants and a report was published in early 2012 (Ministry of Education, 2012, *Mid-Term Evaluation of the School Sector Reform Program*). In this report, the authors proposed to create a renewed focus on achieving quality education (p. 47).
3 Each of the 75 districts of the country is divided into resource centres (RCs) and one resource person (RP) is appointed to supervise all educational affairs within the centre.

References

Akyeampong, K., Pryor, J. and Ampiah, J.G. (2006) "A vision of successful schooling: Ghanaian teachers' understandings of learning, teaching and assessment" *Comparative Education*, 42: 155–176.

Akyeampong, K., Kussier, K., Pryor, J. and Westbrook, J. (2013) "Improving teaching and learning of basic maths and reading in Africa: Does teacher preparation count?" *International Journal of Educational Development*, 33: 272–282.

Alexander, R. (2001) *Culture and pedagogy: International comparison in primary education* Oxford: Blackwell.

Alexander, R. (2008) *Education for all, the quality imperative and the problem of pedagogy* CREATE Research Monograph No. 20, Brighton: University of Sussex.

Alexander, R. (2015) "Teaching and learning for all? The quality imperative revisited" *International Journal of Educational Development*, 40: 250–258.

Archer, D. (2014) "Critical reflections on the learning metrics task force". http://www.actionaid.org/2014/02/critical-reflections-learning-metrics-task-force (accessed 10 December 2016).

Barrett, A. M. (2011) "A Millennium Learning Goal for education post-2015: A question of outcomes or processes" *Comparative Education*, 47: 119–133.

Barrett, A. M. (2016) "Measuring learning outcomes and education for sustainable development: The new education development goal". In Smith, W. (ed.), *The global testing culture: Shaping education policy, perceptions, and practice* Oxford: Symposium Books.

Bruner, J. S. (1996) *The culture of education* Cambridge, MA: Harvard University Press.

CDC (Curriculum Development Centre) (2011) *Nirantar bidhyarthi mulyankan karyanwayan pustika 2068* [*Continuous students assessment: Implementation book*]. Sanothimi: Bhaktapur, CDC.

Dubeck, M. M. and Gove, A. (2015) "The early grade reading assessment (EGRA): Its theoretical foundation, purpose, and limitations" *International Journal of Educational Development*, 40: 315–322.

Hares L. (2013) "Introducing formative assessment in Malawi: Challenges and opportunities". http://sussexciejournal.wordpress.com/2013/10/31/introducing-formative-assessment-in-malawi-challenges-and-opportunities/ (accessed 4 August 2017).

Hattie, J. and Timperley, H. (2007) "The power of feedback" *Review of Education Research*, 77(1): 81–112.

Kapambwe, W. M. (2010) "The implementation of school based continuous assessment (CA) in Zambia" *Educational Research and Reviews*, 15: 99–107.

Lewin, K. (2007) "Diversity in convergence: Access to education for all" *Compare*, 37: 577–599.

Lewin, K. and Stuart, J. (2003) "Insights into the policy and practice of teacher education in low-income countries: The multi-site teacher education research project" *British Educational Research Journal*, 29: 691–707.

Lumadi, M. W. (2011) "Continuous assessment in schools: Teachers' bitter pill to swallow" *Journal of International Education Research*, 7: 27–34.

Ministry of Education (2012) *Mid-term evaluation of the school sector reform program.* Kathmandu: Government of Nepal.

Pryor J. and Kubisi, C. (2002) "Reconceptualising educational assessment in South Africa: Testing times for teachers" *International Journal of Educational Development*, 22: 673–686.

O'Sullivan, M. (2004) "The reconceptualization of learner-centred approaches: A Namibian case study" *International Journal of Educational Development*, 24: 585–602.

O'Sullivan, M. (2006) "Lesson observation and quality in primary education as contextual teaching and learning processes" *International Journal of Educational Development*, 26: 246–260.

Research Centre for Educational Innovation and Development (2004) *Effective classroom teaching learning phase III: School based assessment* Kathmandu: Tribhuvan University.

Schweisfurth, M. (2011) "Learner-centred education in developing country contexts: From solution to problem?" *International Journal of Educational Development*, 31: 425–432.

Somerset, A. (2003) *Basic number skills: Why students fail in math – A diagnostic survey of fifteen high schools in Central Visayas, Philippines* Dilliman: University of the Philippines, National Institute for Science and Mathematics Education Development.

Somerset, A. (2011) "Access, cost and quality: Tensions in the development of primary education in Kenya" *Journal of Education Policy* 26: 482–497.

Somerset, A. (2016) "Questioning across the spectrum: Pedagogy, selection examinations and assessment systems in low-oncome countries". In Smith, W. (ed.), *The global testing culture: Shaping education policy, perceptions, and practice* Oxford: Symposium Books.

UNESCO (2014) *Education for all global monitoring report 2013/4: Teaching and learning* Paris UNESCO.

UNESCO, UIS and Centre for Universal Education (2013) "Toward universal learning: Recommendations from the Learning Metrics Task Force" http://uis.unesco.org/sites/default/files/documents/toward-universal-learning-recommendations-from-the-learning-metrics-task-force-summary-report-2013-en_0.pdf (accessed 10 January 2017).

United Nations (2015) "Transforming our world: The 2030 agenda for sustainable development". https://sustainabledevelopment.un.org/post2015/transformingourworld/publication (accessed 10 January 2017).

Westbrook, J., Durrani, N., Brown, R., Orr, D., Pryor, J., Boddy, J. and Salvi, F. (2013) *Pedagogy, curriculum, teaching practices and teacher education in developing countries* London: Department for International Development (DFID), University of Sussex.

World Education Forum (2000) "The EFA 2000 country assessment report: Nepal". http://unesdoc.unesco.org/images/0012/001215/121535eo.pdf (accessed 18 January 2017).

PART III

Numeracy and power

Facilitating learning of numeracy as social practice

Kara Jackson

The four chapters in this Part illustrate the value of using a numeracy as social practice perspective to understand (and in some cases, design for) the learning of numeracy practices, especially for marginalised groups of people. Each chapter illustrates that numeracy practices are ideological and highlights the crucial role that power plays in making sense of what is or can be learned numeracy-wise, by whom, and for what purposes.

In Chapter 10, drawing on a historical analysis of mathematics education in apartheid South Africa (1948–1994), Khuzwayo provides a compelling account of how mathematics education was used to 'occupy' the minds of Blacks. Through interviews with key figures in the Black education system during apartheid and document analysis, Khuzwayo reveals the ways in which apartheid-era mathematics education was racialised and indeed racist. Blacks were explicitly positioned as incapable of engaging in disciplinary mathematics, and the mathematics curriculum used in Black education at best supported youth to learn ready-made techniques for solving sets of predictable problems. In contrast to other curricula (e.g., history), the mathematics curriculum was largely untouched until recently given the widely accepted conception that mathematics is 'neutral' and 'culture-free'. Against this background, Khuzwayo offers a critique of current mathematics education in post-apartheid South Africa and comments on what it might take to deliberately end occupation of minds through mathematics teaching.

Like Khuzwayo, in Chapter 11, Rampal focuses on the role of the mathematics curriculum in terms of what it assumes about the intended learner and mathematics – and again illustrates the ideological nature of mathematics curricula. Rampal highlights the influence of global and national discourses and policies regarding numeracy, education and assessment on India's construction of mathematics textbooks. Rampal, like Khuzwayo, is critical of current discourses, policies and textbooks; however, the bulk of her chapter focuses on a national effort in which curriculum developers resisted the trend toward 'autonomous' models of numeracy. The resulting national primary textbooks of this era reflect a numeracy as social practice perspective in part because, as Rampal argues, a National Literacy Campaign in the 1990s took seriously the concept of literacy as social practice. As she writes, the textbook developers 'attempted to reframe the social imaginaries of mathematics', that is, who can participate in mathematics and what counts as doing mathematics, through crafting tasks and activities that featured a broader range of people engaged in a broader set of numeracy practices. The analysis provides a vision of what is possible, knowing that texts that aim to reframe social imaginaries need to be accompanied by pedagogy that does the same.

Whereas Khuzwayo and Rampal focus on compulsory education, in Chapter 12, Lekoko and colleagues focus on adult education. They report on a 'transformative pedagogical model for teaching numeracy' that is associated with the Botswana National Literacy Programme. This model is informed by ethnomathematics and critical humanistic perspectives on learning, and privileges a pedagogy that supports learners to 'see' the mathematics that are embedded in their daily practices. Learners are then encouraged to connect their local mathematics practices to more 'formal', disciplinary concepts and procedures. Lekoko and colleagues highlight the potential value in supporting learners – especially those who have either been pushed out or opted out of formal schooling – in coming to see themselves as doers of mathematics in their everyday lives.

Finally in Chapter 13, Yasukawa focuses on yet another context – the workplace – as a site for developing what she calls 'critical numeracy practice', or 'numeracy practices that enable workers to understand, question or challenge the power relations in their workplace'. Informed by third-generation cultural-historical activity theory, Yasukawa reports on two studies of workplace learning in Australia: casual academics (or teaching staff on hourly-paid contracts) in higher education, and workers in firms transitioning to workplace practices aimed at increasing productivity while decreasing costs. She illustrates that changes in both activity systems resulted in new numeracy practices; however the extent to which new numeracy practices resulted in productive outcomes for workers depended on their development of 'negotiating knowledge for the workplace'.

10

'OCCUPATION OF OUR MINDS'

A metaphor to explain mathematics education in South Africa in the apartheid era

Herbert Khuzwayo

In this chapter, I apply Fasheh's concept of occupation of our minds to mathematics education during the apartheid era. I argue that the way South African society was organised during apartheid disadvantaged mathematics instruction for Blacks. As such, redress in mathematics education requires serious commitment to ending occupation of our minds. I am using the concept of occupation in this chapter for two reasons. First, I find it both an interesting and fascinating notion as it directly speaks to dominant Western conceptions of development, of modernity and life as such, which have remained relatively unchallenged by scholars from under-developed contexts. Second, I have found it to be relevant in providing me with answers to some of the questions I grappled with in my study of the history of mathematics education in South Africa from 1948 to 1994.

Munir Fasheh (1996), a Palestinian, coined the term 'occupation of our minds' as a result of his critical look at the Israeli–Palestinian conflict and the effect it had on Palestinian students. In writing about occupation of our minds, Fasheh acknowledges the role that Palestinians have always played in waging war against the occupation of their land and resources. He is, however, critical of the Palestinian curriculum which he claims has been 'meaningless and not built on aspects and issues of the Palestinians' reality' (Fasheh 1997, 27).

According to Fasheh, the struggle in Palestine largely focused on the occupation of Palestinian land and not enough was done to resist occupation of Palestinian minds. In fact, he writes that the biggest danger Palestinians currently face is

> the confiscation of our last possessions as people: our history, our voice, our experience, our vision, our hopes, our unity, our sense of belonging, our rights, our ability to learn and create and our means of survival.
>
> *(Fasheh 1996, 14)*

Fasheh sees the struggle for ending the occupation of our minds as important, since 'the most potent weapon in the hands of the oppressor is the mind of the oppressed' (14). What most characterises the occupation, according to Fasheh, is the following:

> First, the belief that Western cultures are superior to all others, and that the path followed by Western nations was the only path to be followed by others; hence the belief that knowledge and solutions can only come from the West via experts, plans, etc ... Second, preventing our voices, histories and ways of living, thinking and interacting with one another and with nature from surviving and flourishing.
>
> *(Fasheh 1996, 14)*

Fasheh's use of occupation emanates from the fact that Palestinians have been living under Israeli occupation of their land and resources for a long time. In addition to this focus on occupation, Fasheh is concerned that the struggle waged by the Palestinians has not focused on science, mathematics, technology and research. A view in many Western countries is that mathematics is neutral and value-free and that in school it should likewise be taught in a value-neutral way. Fasheh writes that:

> Palestinians like most other peoples in the Third World have been critical of almost everything related to the Western domination except science, math, technology and research. We have considered them an ideal to be critically imitated and followed.
>
> *(Fasheh 1996, 15)*

Fasheh raises questions about the silence of mathematics. Part of the silence of mathematics has come about because of its failure to incorporate broader societal issues as experienced by Palestinians: 'the maths we teach and study, at least in the schools and universities in Palestine is basically like a corpse that doesn't feel anything of its surroundings' (Fasheh 1997, 24).

Fasheh's idea of occupation can be linked to the work and writings of Paulo Freire in the late 1950s and 1960s. Freire was concerned with the development of radical pedagogy, which contributed to progressive social change. He was convinced that to achieve freedom, people must actively wage war and struggle against those stereotypes made of them by their oppressors. Freire further criticised traditional narrative forms of education as oppressive and likened them to a system of banking. He states that education which follows this mode:

> becomes an act of depositing in which the students are the depositories and the teacher is the depositor. Instead of communication, the teacher issues communiqués and 'makes deposits' which the students patiently receive, memorise and repeat. This is the 'banking' concept of education

in which the scope of action allowed the students extends only as far as receiving, filing and storing deposits.

(Freire 1985, 53)

The banking concept of education encourages the form of teaching in which a one-way dependence from the student to the teacher exists. The memorising and regurgitation of facts are important characteristics of banking education. Freire argues that such a process is anti-dialogical and therefore anti-educational on the grounds that dependency presents a contradiction and an obstacle to 'authentic free thinking and real consciousness' (Freire 1985, 85).

Freire (in Gutstein 2006, 88–89) warns against a situation in which 'teachers "deposit" dead morsels of knowledge into passive and subservient students'. Freire points to the disastrous consequences of banking education for students:

The more students work at storing the deposits entrusted to them, the less they develop the critical consciousness which would result from their intervention in the world as transformers of that world.

(Freire in Gutstein 2006, 89)

Freire's critique of 'banking education' can be seen as an attack on all forms of occupation. In a situation where students are depositories and the teacher is the depositor, lack of diversity and standardised thinking is encouraged. There is definitely no seeing of alternatives. Occupation is, therefore, reinforced. Fasheh points out that the key idea to ending occupation involves *inter alia*:

The need to stress in schools the means that help children learn much more than stressing a ready content put forward by experts who have lost their integrity and their senses.

(Fasheh 1996, 21)

Occupation and its relevance in the history of mathematics education in South Africa during the apartheid era

The system of education in South Africa during apartheid encouraged passiveness, memorisation and rote learning, and discouraged curiosity and critical thinking. This was made evident by the persons I interviewed when I conducted a study of the history of mathematics in South Africa during the apartheid years (1948–1994). These interviews demonstrated to me the extent to which learners were denied the opportunity to do creative work.

South Africa is a country where the disparities in mathematics education represent a history of unjust social arrangement. The 1954 Bantu Education Act formally legalised racially segregated educational facilities for all South Africans. According to this act, education was to be based on the concepts of separate development, a separate Bantu society, and a separate Bantu economy for which both Black pupils and Black teachers were to be prepared. This was not a new concept; rather, the political and ideological influences battling over mathematics have had a long history in South Africa. Hendrik Verwoerd, former prime minister of South Africa, claimed that there was no place for mathematics in the education of a Black child. In the address given before the South African Senate, he said:

> When I have control over Native education, I will reform it so that the Natives will be taught from childhood that equality with Europeans is not for them ... People who believe in equality are not desirable teachers for Natives .... What is the use of teaching the Bantu mathematics when he cannot use it in practice? The idea is quite absurd.
>
> *(Verwoerd 1953, 3575)*

According to Verwoerd and the apartheid regime, the paramount task was to make sure that Blacks were prevented from climbing the social ladder. In fact, mathematics was also used as the 'gatekeeper to participation in the decision-making processes of society' (Volmink 1994, 51).

In an attempt to document the history of mathematics education in South Africa, I conducted a research study entitled, 'Selected views and critical perspectives: An account of mathematics education in South Africa from 1948 to 1994' (Khuzwayo 2000). Both the years 1948 and 1994 are important years in the history of South Africa. 1948 marks the initiation of an aggressive and sustained campaign to entrench apartheid in South African law and society, and 1994 represents the beginning of the dissolution of formal government policy that supported apartheid rule. My study explored the history of South African mathematics education for African students during the apartheid period. It places the history of mathematics education in a broader context by describing the general developments in South Africa during this period. The main question the study seeks to address is: 'How did apartheid ideology and political practice influence how mathematics was taught and learned in the past?'. Much of my work in this study was also influenced by my experience as both a student and a teacher during the apartheid period.

To develop an account of the history of mathematics education in South Africa, I interviewed people who had lived and participated in mathematics education in the period 1948 to 1994. These individuals were carefully selected as key figures I assumed to be knowledgeable through their involvement in teaching situations in the Black education system, set up within the broader framework of apartheid education. They were also selected because they had been educated and exposed to an opposite set of political circumstances. For example, Professor W. T. Kambule, a leading figure and a key interviewee, had taught at schools for Blacks and later at a historically White university, whereas

Professor J. P. Strauss had taught at (White) Afrikaans-medium schools and later at a historically Afrikaans-medium university.

In the interviews, I explored participants' experiences as students and teachers, focusing on their recollections about the teaching and learning of mathematics, the mathematics classroom structure, the availability of educational resources, and the political environment during the apartheid era. A number of issues, such as racism, were raised by the people I interviewed. For example, when asked about Verwoerd's statement which suggested that Blacks were incompetent to do mathematics, all of those I spoke with positioned themselves in opposition to the statement. Not everyone, however, positioned themselves in the same way. For example, Strauss (in Khuzwayo, 2000) expressed his displeasure about it: 'I think it was a wrong political statement … [and] … as a teacher, I never believed in mixing politics with education'. On the other hand, Mpofana (in Khuzwayo, 2000) says of Verwoerd's statement: 'It was a deliberate aim that as few [Blacks] as possible should do mathematics'.

Using the interview data, I concluded, for instance, that it was never the intention of the apartheid government (1948–1994) to develop Blacks in mathematics. Classroom activities involved strict formal teaching, questioning by the teacher, and the class working on exercises from the textbook. There was a direct resemblance to the picture painted by Fasheh:

> The classroom is highly organised and the syllabus is rigid and the textbooks are fixed. There is one correct answer to every question and one meaning for every work and that meaning is fixed for all people for all times. Wrong answers are not tolerated; students are usually punished if they make mistakes. Teachers are also expected to perform according to a certain set of rigid expectations and they are punished if they don't.
>
> *(Fasheh 1982, 7)*

I also reviewed archival records such as those of Mathematics In-Service Training Project (MATIP). MATIP was initiated by Professor D. Vermaak of the University of Orange Free State (UOFS) (a historically White Afrikaans-medium university) and launched in 1980. I personally participated as a high-school mathematics teacher in MATIP's courses and workshops. Its role and intervention in Black education was to 'develop' mathematics teachers that were seen as 'worthy'.

As part of my doctoral study, I also investigated what could be called White research on Black education (see Khuzwayo, 2005). One element in this research was to find out why, according to some statistics, Black students could not grasp mathematics. Naturally such statistics had been produced and interpreted within a racist framework and formulated through a deficit model. So it comes as no surprise that the conclusion was that the cause of the failure of Black students was associated with Blacks themselves. A common element in 'white-on-black' research studies was that they invoke race and culture as determining factors of mathematical capacity. Moreover, they make a particular reading of Black culture as problematic, and do so from a particular White perspective which is taken as the norm (Khuzwayo, 2005). Groenewald illustrates this perspective in the following:

> The problems which Blacks experience should be attributed especially to their being rooted in a traditional outlook on and way of life which dictated the pattern of their lives for centuries and which differ fundamentally from the Western way of life. The world of culture in which the Black child finds himself [sic] has restrictive implications for the actualisation of his intelligence.
>
> *(Groenewald 1976, 62–63)*

> Blacks are retarded as regards visual-perceptual development, that in contrast with Whites, they reveal an inability to report depth perception and to interpret three-dimensionally; that their concept of space differs radically from that of Whites; that they experience problems in perceiving pictures and figures analytically; that they do not have a clear understanding of concepts like circumference, length and width and generally find arithmetical concepts difficult to master.
>
> *(Groenewald 1976, 46)*

Rather than questioning and deconstructing the racialised nature of South African society and mathematics learning, statements such as the ones above seemed to ignore an important fact mentioned by Martin:

> what [racial] gaps do highlight are the adverse conditions under which some children are often forced to learn, the privileged conditions afforded to others, and how forces like racism are used to position students in a racial hierarchy.
>
> *(Martin 2009, 301)*

In presenting some justification for the lack of mastery of some mathematics concepts, students' 'cultural differences' were given. In fact, there was a deliberate attempt not to situate this problem into the larger context of apartheid South Africa. As Skovsmose (2005) indicates, mathematics education operates to both provide and justify certain forms of inclusion or exclusion, thereby serving as a gatekeeper. That is, denying access to participate in mathematics is to determine 'who will move ahead and who will stay behind' (Volmink 1994, 52). In South Africa, there was not only occupation of knowledge about mathematics learning, there was occupation of the research in mathematics education as well.

Studying the history of mathematics education in South Africa convinced me that the curriculum that was used during the apartheid era was largely a technique curriculum that emphasised the procedures and steps linked to those procedures. I see the important function of a technique curriculum as having to do with providing the teacher with the toolbox full of techniques which are drilled in the minds of students and which they have to master in order to be successful in their learning. Teachers in turn became the victims of a system which assigned to them the task of implementing a ready-made curriculum and of testing students on techniques. This problem is highlighted in Julie's detailed account of the production of the South African school mathematics curriculum:

> Mathematics teachers thus receive a syllabus describing goals and aims, some comment on methodological aspects, the content to be covered per year and the evaluation procedure to be followed. The intended curriculum reaches teacher after it has been designed elsewhere. Who designed the curriculum; the process that is being followed and the underlying motivations for the curriculum are unknown to those who must implement it.
>
> *(Julie 1991/1992, 4–5)*

It is clear from the quotation above that there is hardly any space available for allowing people to see the alternatives. For example, Bishop is critical of a technique curriculum, for it:

> cannot help understanding, cannot develop meaning, cannot enable the learners to develop a critical stance either inside or outside mathematics … a technique curriculum cannot therefore educate … For the successful child it is at best training, for the unsuccessful child it is a disaster.
>
> *(Bishop 1988, 8–9)*

In South Africa, like in Palestine, we have fallen victim to a widely accepted conception of mathematics as a science that is always true. We have considered mathematics to be a neutral subject that should be taught and learned blindly and uncritically. This view of mathematics – as a perfect system, as pure, as an infallible tool, if well used – contributes to political control (Borba and Skovsmose 1997). This view was in accordance with apartheid conception of education. Viljoen (in Beard and Morrow 1981, 109) for instance, saw education as: 'a science. It revolves around definition, substantion, logical reasoning, experimentation etc.' Further, as Taylor, Adler, Mazibuko and Magadla write:

> Mathematics not in itself but in the way it is constituted, taught and applied … contributes specifically to cultural, class and gender discrimination and to the authoritarian technocracy which dominates all aspects of life in South Africa.
>
> *(Taylor et al. 1986, 11)*

It comes as no surprise, therefore, that whilst subjects such as history were often criticised for being biased in South Africa, mathematics remained untouched for many years after the dawn of democracy. Although profound changes were made to subjects such as history, geography and languages to rid them of their primarily British character, during the early 1980s mathematics was kept intact. It was only in the mid-1980s that this assertion was challenged. For example, new curriculum initiatives were proposed and the role of mathematics within People's Education for People's Power was investigated (Breen 1986).

Also typical of the technique curriculum is its restrictive approach which denies people the opportunity to see alternatives, and which I regard as the opposite of occupation. Wilkinson (1981), in her study of the problems experienced by pupils in mathematics of standard-5 level, mentions as one of her recommendations that the Department of Education should provide each classroom with a daily teacher's guide preferably linked to a textbook:

> this text book should be accompanied by a day-to-day teachers' guide which provides every possible guidance regarding the teaching method, setting and marking of tests and examination papers, revision and enrichment work, etc.
>
> *(Wilkinson 1981, 125)*

The lack (or even absence) of flexibility allowed to teachers is clear from Wilkinson's suggestion above. Any creative work with the learners is denied, thereby denying the opportunity for both the teachers and students to see alternatives in their work. Perhaps it is possible to find some justification in Wilkinson's suggestion if it is taken into consideration that South Africa has always lacked teachers who are qualified in mathematics and science (see for example, a report for the Department of Education and Training and the Department of Arts, Culture, Science and Technology – Arnott et al. 1997). The problem, however, with the above suggestion stems from my belief that it is highly unlikely that the situation can be improved by further disempowering teachers through a system that encourages passiveness, rote learning and obedience to authority, while also discouraging intellectual risk taking, curiosity or independence of thought.

Wilkinson's suggestion was strictly adhered to by the Department of Education. (I am, however, not suggesting that this came about as a direct result of Wilkinson's suggestion). I am convinced, though, that this tendency, as reflected in Wilkinson's suggestion, in turn severely limited the options and alternatives in peoples' minds. It is about blocking learners' options, thereby implying that there is only one way of doing things, emphasising strict algorithmic approaches which deny learners' creativity in learning mathematics. Most textbooks during the time of apartheid education resembled the description given by Volmink:

> Most school textbooks are written in a style which emphasises drill and practice or routine exercises. At the end of these exercises, some

space is given for problems for which a standard recipe for obtaining the answer cannot be used. These problems are generally decontextualised. At best, they are applications of a previously learnt principle or concept. So application problems in school textbooks are used mainly to provide exercises in or to illustrate one or other mathematical technique.

(Volmink 1994, 61)

The majority of teachers in the White education system had also come out of an education system which was based on Fundamental Pedagogics (FP), an authoritarian philosophy associated with Christian National Education and apartheid (Enslin 1990). Perhaps the most serious charge that can be laid at the door of FP, which dominated apartheid education, is that it discouraged the very qualities regarded as essential for sustainable development and success as the new millennium approached: risk taking, a sense of adventure, curiosity, a critical and questioning attitude, self-motivation and reflection, inventiveness and independence of mind; in a phrase, creativity and innovation (Arnott et al. 1997, 6). In contrast, South Africa had a system of education that discouraged independent thought. Parker (in Eshak, 1987) says that in FP the child has to submit to the authority of the teacher and the teacher to a set of norms laid down by a higher being. The individual must submit to the authority of the state which represents the higher being. For the individual to act freely, the individual must act in accordance with the laws of the state. The stress on obedience makes education a process of submission. This was in accordance with a view held by the proponents of FP, which viewed education as a science based on definitions, logical reasoning, experimentation, etc.

The proponents of FP (see, for example, Viljoen and Pienaar 1971) also saw the role of the teacher as authoritarian. Gunter notes, for instance, that in FP:

the educator is invested with authority and as such he [sic] has the right to prescribe to the educand what he must do, and how or what he must not do, while the educand has to respond to his being addressed by the educator by accepting what he says.

(Gunter 1974, 144)

FP was a powerful weapon used by apartheid ideologues to occupy not only the minds of Blacks but also Whites. This must not come as a surprise, as mathematics learning and participation is 'structured by the relations of race that exist in the larger society' (Martin 2009, 299). This relationship is not limited solely to South Africa, but may be seen in other contexts, as well. For example, when writing about the race problem in mathematics education in the US, Martin maintains that, 'all students – not just those identified as African American, Latino, and Native Americans – experience mathematics learning as a racialised endeavour' (Martin 2009, 299–300).

The statements above indicate that the way South African society was organised during apartheid contributed significantly towards disadvantage in mathematics for Blacks.

Mathematics both ways: an example of a means to end occupation of our minds

How can we work to end the occupation of our minds in mathematics? Stanton's (1994) Australian example of 'Mathematics both ways: A mathematics curriculum for Aboriginal teacher education students' provides some insight. The struggle for land by Aboriginal populations in the Northern Territory of Australia as a result of colonisation of original inhabitants by White settlers is similar in many respects to the one which has been faced by the Palestinians. Lanhupuy (in Stanton 1994, 15) argued that schools have been part of the colonisation and dominance which has resulted in 'influencing the minds of children in ways that undervalue the Aboriginal heritage'. Aboriginal people have been waging a struggle to free themselves from occupation of the mind as they now 'understand that if schools are to serve the political, social and economic purposes of their own people, the schools need to be accommodated within Aboriginal culture itself' (Lanhupuy in Stanton 1994, 15). As part of the solution to freeing the minds of children, they have embraced 'both ways' education. The notion of a 'both ways' education is described by Stanton as 'an education that recognises the validity of the knowledge bases from the Western and Aboriginal traditions' (Lanhupuy in Stanton 1994, 15). This contrasts with the curriculum of White schools in which the focus is on one-way Western traditions. In an attempt to decolonise their schools, the Remote Area Teacher Education's (RATE) 'both ways' pedagogy places emphasis on problem-posing/problem-solving approaches to learning, curriculum negotiation and integrated curriculum planning supported by appropriate assessment strategies including criterion referencing, descriptive reporting and non-competitive assessment. It is community-based and community-focused, and aims to have a role in developing its students' skills in the defence, maintenance and further development of Aboriginal culture.

'Both ways' education does not see mathematics as the mastery of classroom techniques which have no real-world relevance and are owned by the dominant culture. The most obvious way in which 'both ways' mathematics curriculum looks different from the standard techniques-oriented curriculum is that it does not consist of a list of techniques, sequenced in terms of an arbitrary hierarchical structure. Instead, the techniques may be found subsumed under the notion of six-component symbolic activities that Bishop (1988, 19) proposes are found across all cultures. These symbolic activities are 1) counting, 2) locating, 3) measuring, 4) designing, 5) planning and 6) explaining. Stage one of the 'both ways' course followed at Bachelor College of Teacher Education, for instance, has as its central theme the social contexts of mathematics learning. The focus is on 'What is this thing called mathematics?' and 'How is mathematics used in my community?' (Stanton 1994, 19). Other curriculum activities centre on:

> ways which demystify and make mathematics accessible to the Aboriginal teacher and child alike, ways which allow the Aboriginal community to co-opt mathematics, its symbolic technology and machines.
>
> *(Stanton 1994, 19)*

Though it is not possible to describe everything in this chapter about the mathematics curriculum at the Bachelor College of Teacher Education, an attempt to end occupation is realised. Through the 'both ways' education approach, opportunities are created for students to see the alternatives. The mathematics curriculum which students use is not imposed from above but is negotiated between lecturer and student. Most importantly, Aboriginal students are able to deal with Western mathematics, described by Bishop (in Stanton 1994) as one of the most powerful weapons in the imposition of Western culture. Lecturer and students, for example, engage in activities that provide focus on issues for community research that have implications for the development of Aboriginal mathematics pedagogy (Stanton 1994).

Are current reforms in mathematics teaching adequately addressing the problem of occupation in South Africa?

I believe that one of the important challenges facing all those who are involved in mathematics education (and education generally) in South Africa today, is to begin to explore ways and means of ending the occupation of our minds brought about by years of apartheid education. The challenges facing curriculum developers in South Africa are particularly daunting, given the heritage of passivity, prejudice, ignorance and resistance. Fasheh has an important message for those who are committed to ending the occupation:

> Ending the occupation of our minds is a personal task; its continuation depends solely on our acceptance of it. So is its termination. Since ending it is crucial for ending other forms of occupation, and for building our future, it is a main challenge.
>
> *(Fasheh 1996, 25-26)*

Since the onset of democracy in 1994, the South African government has introduced a series of curricular initiatives designed to effect quality education for all learners. As South Africa moves forward with new curricular initiatives that are aimed at the elimination of many of the disparities of the past, I believe the education directed at ending our occupation should be an important consideration. My only fear is that the kind of progressiveness which has motivated some of these curricular changes has not benefitted most teachers at the classroom level. I see this resulting from the unfortunate lack of debate prior to the implementation of such changes, like the outcomes-based curriculum.

What we have continued to witness instead is that the reform processes have been left to 'experts'. Most curricular changes have been prescriptive and usually monitored by government officials. Ramatlapana and Makonye (2012) mentioned that the latest curricular transition from National Curriculum Statement (NCS) to Curriculum and Assessment Policy Statement (CAPS) has compromised educator autonomy. The authors further note that the CAPS curriculum is prescriptive to the point of demanding uniformity in implementation across the nation. Prescriptive curriculum is unhelpful to teachers as it continues to block them from seeing alternatives that help them contribute meaningfully to the process of educational change. The CAPS curriculum, therefore, is not freeing students and teachers from occupation of the mind as it continues to promote a teacher-centred pedagogy.

Conclusion: ending occupation

Freeing people from occupation is both a worthwhile and a just cause. If we are concerned about the education of our children, we need to think seriously about acquiring the means to end occupation. This should begin with the realisation that a threat posed by occupation always exists for both the victim and the perpetrator. An important consideration should be to allow people to see the alternatives and to re-contextualise knowledge. This is an important message for South Africa. There is no doubt about the extent of damage caused by many years of apartheid education in South Africa.

Fasheh's notion of occupation of our minds is not only relevant to our past and present; it has an important message for our attempts to transform mathematics education in South Africa for the future – and for the better. Any meaningful curricular reform requires that we start by:

> shaking off the dirt that has been accumulated in our minds, mainly through formal education (including science), television and other killers of cultural diversity and of human societies.
>
> *(Fasheh 1996, 19–20)*

In South Africa, we need to think carefully about how we are going to be 'shaking off the dirt that has been accumulated in our minds' which has been left by long years of FP which dominated apartheid education. This can be achieved in more than one way. We can, for instance, do this by incorporating some ideas of 'planting the seeds' (Fasheh 1996, 23) mentioned by Fasheh and also incorporating some elements of the 'RATE both ways: Curriculum for Aboriginal teacher education students' model (see Stanton; referred to earlier in this chapter). The kind of 'seeds' Fasheh mentions are, for instance, the following:

- recognising and supporting teachers who use innovative ways of teaching mathematics;
- recognising and supporting students who are involved in some interesting problems;
- integrating the teaching of art and mathematics;
- integrating the teaching of language and mathematics;
- producing materials that could improve the learning process in the traditional classroom;
- holding a 'contest' to encourage teachers (as individuals or groups) to write inspiring and relevant questions that relate mathematics to local conditions, etc.

We should consider what is suited and relevant for the needs of South Africa. Most importantly, we should guard against importing. For our curriculum reforms to have meaning, they should be based on aspects and issues of the South African reality. We should be careful that our endeavours to end occupation are carefully considered. The greatest mistake we continue to commit is to replace apartheid-type of occupation by other forms of occupation, including having a curriculum that is top-down and 'expert'-driven. Importing a curriculum with disregard to existing realities of our situation would not only result in a waste of our limited resources, but would also be disastrous for the new South Africa.

In ending occupation, in Khuzwayo (2016), I have called for the adoption of *Ubuntu*, the underlying philosophy that shapes and drives African intellectual interests and his/her way of life as a whole in the teaching and learning of mathematics. *Ubuntu* emphasises communality rather than individualism, prevalent in most Western societies. Nussbaum (2003, 21) describes *Ubuntu* as the capacity in African culture to express compassion, reciprocity, dignity, harmony, and humanity in the interest of building and maintaining humanity. *Ubuntu* is based on such principles as encouraging the relationship of people to people rather than people to things as is the case in modern technological societies. *Ubuntu* accommodates different views and opinions, and standardised thinking is discouraged. Embracing *Ubuntu* to end occupation would require paying particular attention to recontextualising knowledge, developing mathematical understanding, and providing support to teachers. Integrating *Ubuntu* in the teaching of mathematics will assist the teachers to do away with teaching it as pure and detached from the context in which people live. Whilst there might be a need for a core curriculum, we must guard against a situation where teachers are discouraged to use professional judgment in their work.

References

Arnott, A. Z., Kubeka, A., Rice, M. and Hall, G. (1997) *Mathematics and Science Teachers: Demand, Utilisation, Supply and Training in South Africa* Craighall, South Africa: EduSource.

Beard, P. and Morrow, W. (1981) *Problems of Pedagogics* Pretoria: Butterworths.

Bishop, A. (1988) *Mathematical Enculturation: A Cultural Perspective in Mathematics Education* Dordrecht: Reidel Publishing Co.

Borba, M. C. and Skovsmose, O. (1997) "The ideology of certainty in mathematics education" *For the Learning of Mathematics*, 17(3): 17–23.

Breen, C. (1986) "Alternative mathematics worksheets" unpublished paper University of Cape Town.

Enslin, P. (1990) "Science and doctrine: theoretical discourse in African teacher education". In Nkomo, M. (ed.), *Pedagogy of Domination* Trenton, NJ: Africa World Press.

Eshak, Y. I. (1987) "'Authority' in Christian national education and fundamental pedagogics" a research report submitted to the Faculty of Education, University of the Witwatersrand, Johannesburg, in part-fulfilment of the requirements of the degree of Master of Education.

Fasheh, M. (1982) "Mathematics, culture and authority" *For the Learning of Mathematics*, 3(2): 2–8.

Fasheh, M. (1996) "The main challenge: Ending the occupation of our minds. The main means: Building learning environments and re-contextual knowledge". In *Political Dimensions of Mathematics Education III*. Bergen: Caspar Forlag.

Fasheh, M. (1997) "Is math in the classroom neutral of dead? A view from Palestine" *For the Learning of Mathematics*, 17(2): 24–27.

Freire, P. (1985) *Pedagogy of the Oppressed* Harmondsworth: Penguin.

Groenewald, F. P. (1976) *Aspects in the Traditional World of Culture of the Black Child which Hamper the Actualization of his Intelligence: A Cultural-Educational Exploratory Study* Pretoria: South African Human Sciences Research Council.

Gunter, C. F. G. (1974) *Aspects of Educational Theory: (With Special Reference to the School)* Stellenbosch: University Publishers and Booksellers.

Gutstein, E. (2006) *Reading and Writing the World with Mathematics: Toward a Pedagogy for Social Justice* London: Taylor & Francis.

Julie, C. (1991/1992) "Equation of inequality: Challenging the school mathematics curriculum" *Perspective in Education*, 13(1): 3–10.

Khuzwayo, H. B. (2000) "Selected views and critical perspectives: An account of mathematical education in South Africa from 1948–1994" doctoral dissertation, Institute of Electronics Systems, Department of Mathematics and Computer Science, Aalborg University, Denmark.

Khuzwayo, H.B. (2005) "A history of mathematics education research in South Africa: The apartheid years" Vithal, R, Adler, J and Ketiel, C. (eds) *Researching Mathematics Education in South Africa: Perspectives, Practices and Possibilities* Cape Town: HSRC.

Khuzwayo, H.B. (2016) "Ending the occupation of our minds: The adoption of Ubuntu in the teaching and learning of mathematics and science". In Marcos, C. (ed.), *Proceedings of the 2nd International Conference of the African Association for the Study of Indigenous Knowledge Systems,* Universidade Pedagogica, Maputo, 27–29 October. Cape Town: University of the Western Cape.

Martin, D.B. (2009) "Researching race in mathematics education" *Teachers College Record*, 111(2): 295–338.

Nussbaum, B. (2003) "Ubuntu: Reflections of a South African on our common humanity" *Reflections: The Sol Journal*, 4(4): 21–26.

Ramatapana, K. and Makonye, J. P. (2012) "From too much freedom to too much restriction: The case of teacher autonomy from National Curriculum Statement (NCS) to Curriculum and Assessment Statement (CAPS)" *Africa Education Review*, 9(1): S7–S25.

Skovsmose, O. (2005) "Foregrounds and politics of learning obstacles" *For the Learning of Mathematics*, 25(1): 4–10.

Stanton, R. (1994) "Mathematics both ways: Mathematics curriculum for Aboriginal teacher education students" *For the Learning of Mathematics*, 14(3): 15–23.

Taylor, N., Adler, J., Mazibuko, T. and Magadla, L. (1989) "People's education and the role of mathematics" conference workshop organised by the Department of Education, University of Witwatersrand, Kenton-on-the-Juskin, 31 October–3 November.

Verwoerd, H. F. (1953) *House of Assembly Debates*, 78 3575–3670.

Viljoen, T. and Pienaar, J. (1971) *Fundamental Pedagogics* Durban: Butterworth Publishers.

Volmink, J. (1994) "Mathematics by all". In Lerman, S. (ed.), *Cultural Perspective on the Mathematics Classroom* Dordrecht: Kluwer Academic Publishers.

Wilkinson, A. C. (1981) "An analysis of the problems experienced by pupils in mathematics at standard 5 level in the developing states in the South African context" M.Ed. dissertation, University of the Orange Free State, Bloemfontein, South Africa.

11

LEARNING FOR LIFE, FROM LIFE

Adult numeracy and primary school textbooks in India

Anita Rampal

Introduction

Global economic pressures for standardisation of school mathematics, tied to aggressive international testing and congruent national assessments, have been eroding the space that the World Declaration for Education for All, 1990, had afforded in diverse developing countries such as India. It had called for an 'expanded vision for education' (UNESCO 1990, 3), 'recognising that traditional knowledge and indigenous cultural heritage have value and validity in their own right and capacity to both define and promote development' (UNESCO 1990, 2). In India, this had aligned well with the participatory national campaign for adult literacy (and numeracy) in the 1990s, addressing critical agency through cultural mobilisation with a focus on local and indigenous knowledge. Though the global discourse changed over the next decade, towards more instrumental aims and 'learning outcomes', it was resisted nationally by an unprecedented alignment of academics and social activists who came together to restructure school education under the umbrella of the National Curriculum Framework 2005 (NCERT 2005). Many of these persons had engaged with curricular issues earlier as part of collaborative innovative programmes in the states, such as Madhya Pradesh, Rajasthan, Kerala, Karnataka, etc. Contributing at the national level under a liberal political dispensation gave this collective effort significant voice to reframe the syllabi and textbooks of the National Council of Educational Research and Training (NCERT), bringing together diverse experiences from the field. The National Curriculum Framework 2005 espoused a social constructivist approach, which focused on learners, their agency and the social and cultural rootedness of processes of meaning making and knowledge construction. There were major creative shifts in the social science curricula, including history and political science at the secondary stage, and also in the primary curricula for language, mathematics and environmental studies.

This chapter looks at the links between the development of the current national primary mathematics textbooks (first published by NCERT in 2006–8), and the National Literacy Campaign in the 1990s, which had engaged with numeracy as social practice, interwoven with people's lives, their contexts and aspirations, historically and culturally. This is in sharp contrast to the earlier adult education programme run by the government prior to the 1990s which reflected a deficit view, focusing on adult learners' 'inability' to deal with mathematics, and doled out a minimalist content of childish procedural arithmetic. The participative nature of the National Literacy Campaign afforded an exploratory way of critically understanding adult learners' complex everyday practices, which resulted in curricular materials and tasks designed to challenge and motivate adult learners. By the early 2000s the campaign was stalling, owing to increased bureaucratic demands and lack of congruent political and financial support. Yet, in a way, its legacy lived on and helped shape a culturally responsive pedagogy of mathematics for the national primary school textbooks.

As the chairperson of the NCERT Textbook Development Teams for the Primary Stage, I was able to invoke our earlier experiences of working with the Literacy Campaign and incorporate people's funds of knowledge (Civil 2007) to re-envision children's textbooks (see *Math-Magic* for classes 2–5, published by the National Council for Educational Research and Training, available online at http://epathshala.nic.in/e-pathshala-4/flipbook/). Tying quality to equity, which requires high expectations from and opportunities for *all* children to perform well (Boaler 2008; Nasir and Cobb 2007), the challenge of 'math for all' was addressed by restructuring the nature of primary mathematics. In a conscious effort to bridge the mathematical discourses of the home, community, and school, these textbooks included narratives of real-life protagonists doing situated mathematics to resist the valorisation of certain privileged knowledges within the hierarchies of school subjects (Yasukawa and Black 2016). These textbooks also attempted to reframe the social imaginaries about mathematics, about who could and could not participate, and why, so that change was talked into being (Hamilton 2016; Barton and Hamilton 2012) through personal stories of everyday agency with which students could relate.

The National Curriculum Framework 2005 was ratified by the Right to Education Act (Government of India 2009) for children aged 6–14 years. This historic entitlement came after a long struggle by civil society groups. Section 29 (2) of the Act mandates a curriculum for learning through activities, discovery and exploration in a child-friendly manner, making the child free of fear, trauma and anxiety. The Right to Education Act banned any selection of children at the time of admission and mandated that they could not be denied promotion to the next year before they complete elementary school at Class 8. It recognised their right to learn, and held the system responsible for failing to ensure that. This is most significant for mathematics, which continues as a 'killer' subject, causing a majority of children to suffer from anxiety and loss of confidence, and resulting in many students routinely labeled as 'slow' or 'stupid' and ultimately being 'pushed out' when they fail.

The present economic 'skills' discourse is contrary to the egalitarian national vision of 'work in education', as historically established by the philosophy of 'Nai Taleem' (literally, new knowledge). The Gandhian system of schooling (Sangh 1938), part of the anti-colonial struggle for freedom, called for 'education for life, through life'. These schools set up in the 1940s and early 1950s also sought to generate funds through productive work in order to be independent of scarce government finances. A productive craft – such as weaving, carpentry, agriculture, or pottery – was located as the 'medium' of learning, to integrate the head, heart and hand. It is significant to note the similarity between the relationship of 'knowledge' and 'work' as seen in Nai Taleem, and contemporary theories of 'situated learning' (Lave 1996) that informed our efforts in adult numeracy. Nai Taleem was based on the belief that 'knowledge comes through work, that work is inspired by knowledge, and that life is made fruitful by the experience of work and knowledge as one' (Vinoba 2014, 63). More importantly, by locating 'work' (including cleaning the school premises and toilets) at the centre, Nai Taleem interrogated the entrenched caste system that socio-economically stigmatised the 'low-castes' and their vocations as 'untouchable'. A few schools still follow the Gandhian legacy, and in their own way challenge the now aggressively adopted neoliberal agenda of a changing education system, which in fact reinforces social inequities by forging other divides of knowledge/skill, academic/vocational, English-medium/mother tongue, and so on.

The National Curriculum Framework 2005 mandates a common curriculum for all until Class X, with vocational options at the higher secondary stage, but recently there have been moves by the central and state governments to segregate children through vocational and 'distance learning' courses much earlier, on the basis of their 'low achievement'. Conventional school mathematics and assessment practices serve such early discrimination and segregation. Concerted pressure from international agencies and corporate advocacy groups, pushing governments into standardised testing, and using learning outcomes to shame public education, narrows the discourse, compelling central and state governments to abandon larger visions of education for equity and cultural rootedness of learning. In keeping with this agenda, the present government has revoked important aspects of the Right to Education and has allowed states to not promote children even at the primary stage, in its push for more aggressive testing and sorting. With more than half our children not able to complete secondary school, the Board examinations (at Class X and XII) serve to support the hegemony of mathematics as labeller and gatekeeper for future life trajectories. In a poor country with vast inequalities, the myth of mathematics only for the talented helps to reserve future opportunities and privileges, including coveted fields of higher education, for those who can buy their way in, through a burgeoning market of private coaching, guidebooks and information technologies.

Ironically, just when the Right to Education Act was in the process of finalisation by the central Ministry of Human Resource Development, the World Bank had negotiated directly with two Indian state governments and pushed them into taking the Programme of International Student Assessment (PISA) tests. These international tests, conducted every three years by the Organisation for Economic Cooperation and Development (OECD) to compare the achievement of 15-year-olds in language, mathematics and science literacy, dominate the arena of global educational governance and have caused concerns regarding their narrow educational aims that do not recognise the complexity and cultural bases of teaching and learning (Addey 2017, 1). The two Indian states chosen had high levels of enrolment and had striven for basic provisions for public education, despite low resources. Expectedly, pushing poor countries into PISA, where schools are struggling with abysmally inadequate infrastructure and teachers, and do not have the ability to address the demands of literacy or numeracy as posed by such international tests, leads to hysterical chest-beating in the media on their dismal performance, and further shames and demoralises the public system. The Indian government has resisted PISA since 2009, but there are indications that it might soon enter the fray. This 'scandalising' discourse of failure further serves the interests of the privatisation lobby, and boosts the billion dollar market for low-fee private schools, actively promoted in developing countries, with low-paid 'instructors' programmed to use tablets to deliver 'outcomes'.

There are thus increasingly daunting challenges of developing curricula for schools as diverse and increasingly iniquitous as in India, through pedagogies of empathy that enable democratic participation (Rampal 2013). Yet there is all the more reason to pursue this course. Work towards transforming mathematics curricula to ensure 'learning for life, from life', drawing inspiration from the adult numeracy programme, and the egalitarian vision of Nai Taleem decades earlier, continues to contribute to diverse domains of educational practice in India. For instance, the recent national two-year bachelor of education (BEd) programme for teacher education, developed by a committee which I had chaired, for the first time includes aspects of mathematics as social practice and for social justice. Some PhD courses at the University of Delhi were specially designed and led to doctoral research theses on culturally responsive mathematics, which focused on developing curricula for middle school, on assessment of mathematics, teachers' beliefs, and so on. The following sections present some learnings from this long, exciting and very challenging journey.

Adult literacy campaigns: connecting 'learning', 'meaning' and 'agency'

The 1990s witnessed a significant participatory movement in India through large-scale literacy campaigns conducted in over five hundred districts in the country, to involve non-literate adults in the age group of 15–45 years. A voluntary organisation Bharat Gyan Vigyan Samiti (better known as BGVS) was formed to work collaboratively with the National Literacy Mission of the government. The local teams helped decentralise the planning and implementation under the aegis of the district administration. Teaching was carried out by millions of volunteers without remuneration; the government bore a low cost (less than US$2 per learner), while the community pitched in through local organisers, volunteer teachers, space for literacy classes, food and shelter for the travelling 'jatha' or cultural groups engaged in mobilisation and orientation. For the first time in Indian education, the curriculum and primers were meant to be developed at the district level, instead of the state or national levels, and in local languages, including some indigenous languages not scripted earlier. Many districts made primers and teaching and learning materials to focus on themes from their specific local contexts, such as working in coal mines, migration, agriculture, local self-governance, women's self-help groups, water harvesting, etc.

Oral numeracy was more natural among unschooled adults than literacy; they were more familiar with numbers, operations, estimations and measurements, etc. than with alphabets. However, in the traditional primers made by the State Resource Centres, the numeracy content was appended to each lesson in an ad hoc manner. As far as reading and writing were concerned, there seemed to be some understanding in the campaign that teaching would codify speech and verbal thought, through generative words, and not start mechanically from alphabets. However, there was no such attempt to understand the lived 'meaning of numbers' or to look for generative themes for numeracy. In a review of the earlier teaching learning materials for adult learners I had noted that the conventional primers laboured through numbers in a childish and linear fashion, 1–10 in Chapter 1, 11–20 in Chapter 2, and so on (Rampal, 2003). Pictorial illustrations were taken from children's books, and learners were condescendingly asked to count eight ducks or five apples. On the contrary, adults wanted to move on to more sophisticated and challenging tasks in mathematics, to help them deal with market transactions more confidently, and to avoid getting cheated. Adults were capable of doing 'word problems' before they learned to read and write. However, conventional adult educators, influenced by the school myth of 'word problems are more difficult' avoided these as well as other everyday calculations involving profit and loss or simple interest on loans considered important by the learners themselves. Traditional adult educators were also wary of fractions, unaware of how some adults could comfortably negotiate orally through complex fractions as part of their work. In many south Indian states, intricate fractions had continued to exist in everyday vocabulary, with possible links to the markedly intricate sound and beat patterns used in classical Carnatic music. In Tamil (the language of the state of Tamil Nadu) and Malayalam (of the state of Kerala), people spoke of 'half of one fourth' as 'araikaal' or 'three fourths of one eighth' as 'mukaal arakaal'. This, interestingly, is not found in the languages or the music patterns of the north Indian states.

Similarly, 'dichotomous divisibility' or repeated divisibility by two, significant in the counting systems with base 12 or 16 or the measures used by traditional societies could be found in everyday use but not in the primers (Rampal 2003).

Traditional numeracy curriculum development had been devoid of an understanding of how people learn while doing mathematics and the kind of strategies they adopt in diverse social and cultural settings. Real-life learning is not necessarily a lone struggle by an individual working in isolation, but is a collective social process (Daniels 2001). In most unschooled situations, learning takes place as a 'situated activity', grounded in specific transactions, through shared practices where people scaffold each other through zones of proximal development that are collective rather than individual, and attain new knowledge, skills, and strategies through the learning interaction (Lave and Wenger, 1991). Sociocultural theories of learning do not separate thought, action and feelings. Accordingly, knowledge and learning cannot be 'pinned down' to the head of a person, but are situated in the process of work – in the relations between the person, the activity or task being done, the tools used and the setting or the social and physical environment. Knowledge is thus constantly constructed and transformed as we perform an activity. Planning learning activities therefore requires that we know how learners – adults or school children – construct knowledge from their everyday observations, practices and intuitive theories.

We had noted that situated theories (Cobb and Bowers 1999; Gee 2004; McDermott 1996) view 'learning disability' in school as more to do with children having to face more 'arbitrary' rather than 'difficult' tasks at school, for which they are unable to define meaning and purpose, for its delayed use at some probable point in life. A static theory of learning prevails in formal education, relying on a static view of context where:

> the term context refers to an empty slot, a container, into which other things are placed. It is the 'con' that contains the 'text', the bowl that contains the soup …The soup does not shape the bowl, and the bowl most certainly does not alter the substance of the soup. Text and context, soup and bowl … can be analytically separated and studied on their own without doing violence to the complexity of the situation. A static sense of context delivers a stable world.
>
> *(McDermott 1996, 282)*

An 'activity' is shaped by the active contributions from individuals, their social partners, historical traditions, and also materials and their transformations. Unlike the 'storage model' of mind, where static elements of knowledge are believed to be held in the brain at one point of time, and mysteriously processed to be used at some other time in life, 'participatory appropriation' (Rogoff 1995, 150) is a perspective about learning as an ongoing activity. People participate in activities and learn to handle subsequent activities in ways based on their involvement in previous events. There is no need to segment time into past, present and future, or to conceive of learning as 'internalisation' of stored units, achieved after external exposure to knowledge or skills. As a person acts based on previous experience, the past is present in the participation, and contributes to the event by having prepared the ground for it. The present event is therefore different from what it would have been if the previous events had not occurred, and does not depend only on stored memory of past events.

To understand how people constitute learning, it becomes important to look closely at their agency and the processes through which they participate in a cultural activity, as well as how they also transform that process. It is broadly within this perspective that we engaged with the understanding of numeracy in the literacy campaign.

Numeracy Counts!: a handbook for adult learners

As resource persons we had conducted workshops with volunteers and writers of primers and other post-literacy materials, to address the cultural repertoire of adult learners, who though mostly unschooled, were confidently engaged in *doing* mathematics as part of their daily activities. From these interactions, it emerged that we needed to develop a numeracy handbook (Rampal, Ramanujam and Saraswathi 1998; 2000; both accessible online), titled *Numeracy Counts!* in English, and an expanded Hindi version, *Zindagi ka Hisaab,* which could be used by resource persons and volunteers in different states to take their work further.

The handbook located the programme of adult numeracy within a socio-cultural perspective and shared experiences from within the country and elsewhere. Observations of street mathematics in Indian situations showed a similarity with those in Brazil (Nunes et al. 1993), and gave insights into the often sophisticated oral strategies people use. It provided a critique of the conventional teaching approaches and shared diverse resources to help change the way numeracy sessions could be conducted. For instance, it suggested to the volunteer teachers to think of the meaning of numbers in their lives and begin by maintaining a number diary for their Literacy Centre. The volunteer teacher would ask adult learners to make a simple number statement about themselves. For instance: 'I weigh 53 kilograms', 'My uncle has eleven toes', or 'There are seven people in my family'. The volunteer teacher could then initiate a discussion with a series of questions such as:

- How many films do you remember having seen in the last five years?
- How many mangoes can you expect to buy for ten rupees?
- What is the number of the bus you take to go to … from …?
- What is the cost of one kilo of wheat?
- How many stars do you have on your *lehenga* (skirt)?
- How many buckets of water do you use every day?
- How many trees are there in your village?

The volunteer then gave exercises based on the numbers that had arisen out of these discussions. Another exercise involved the volunteer saying a number and asking the learners to respond with some objects of that quantity. For instance, the volunteer would say '100' and a learner may respond with '100 jasmine flowers.' Learners would also be encouraged to ask questions of each other, to look consciously for numbers, and contribute to the growing collective database of familiar numbers. Questions which required estimations of large numbers were deliberately posed:

- How many mangoes does a tree yield in a year?
- How many leaves does a typical mango tree have?
- How many *rotis* do you make in a year? How many do you eat yourself?
- How many stars are there in the sky?
- How many people live in our village?
- How many hairs do you have on your head?

A challenging game took the form of bidding, 'whatever number you tell me, I will tell you a greater number'. Gradually, when numbers got into the thousands, the volunteer teacher could go for much larger increments, and even when they did not know the exact relationships, say, between a thousand and a lakh (hundred thousand), it was enough to know the latter was much bigger. This honed the learners' own estimation skills, as was used for the leasing of fruit trees, where contractors would predict the likely yield of a tamarind or mango tree. Exercises for large estimates were based on a series of Fermi-type questions. We began with questions such as, 'What is the number of cups of tea drunk this morning by the entire village?' and proceeded to discuss strategies to estimate. This further led to interesting group activities of planning, say, for a village feast, with estimates and quantities, and elaborate descriptions of the recipes. Needless to say, volunteer teachers were required to be trained to handle such sessions effectively. Some along with resource persons also conducted participatory resource mapping, in which adult learners estimated the number of children below the age of ten years, the number of cows and buffaloes, and so on, and later graduated to more systematic enquiries, which generated valuable databases of local information. In several places, detailed maps were drawn by the neo-literates themselves, and were used for watershed management programmes for the village. This helped the group develop a critical understanding of their world (Jordan 2012), with the possibility of seeking more control over their lives. Indeed, a neo-literate woman who estimated the number of *rotis* she had cooked in her life suddenly exclaimed that the exercise had changed her perspective about herself, her life and women's work!

The numeracy handbook contained examples from people's repertoire of folk stories, puzzles and riddles about numbers and suggested how volunteers could elicit more from the adult learners, in different regions across the country. It discussed that oral societies have invested effort, ingenuity and agency in devising mnemonic techniques to memorise, preserve, and transmit to future generations their rich bodies of knowledge. *Shlokas, mantras* and *sutras* rendered through elaborate rhythmic patterns were all means to ensure that the rich knowledge of such societies was made memorable for posterity. Moreover, verse and rhyme, which help in memorising long pieces of complicated information, have been woven creatively with the empirical observations and philosophical moorings of oral civilisations (Rampal 1992).

Adult learners have continued to engage in a host of mathematical transactions, such as sorting, measurement, estimation, making patterns, etc. (Gerdes 1985; Greer et al 2009) as part of their activities related to life and labour (Ghose 2007). New ideas or skills provided by the literacy programme needed to be reinterpreted through the learners' own mediatory mechanisms, of assigning meaning to them, and, more importantly, testing them out in real-life settings. However, teaching practices most often do not help learners reflect on the challenges the new measures place on them. As pointed out by economic historians, the traditional measures were truly 'representational', those that 'signified' something, and therefore had a social 'meaning', while it is the metric system that is wholly arbitrary and dependent on convention (Kula 1986). For instance, not many of us might know that the metre was first defined by the French as the '1/40,000,000th part of the earth's meridian', which was then considered to be immutable and unchanging but not now. Subsequently, with the requirement of greater accuracy and reproducibility, the dependence on a singular physical entity was abandoned in favour of a more abstract and almost metaphysical definition. Thus, the way a metre is defined today can afford very little 'meaning' to us – namely, the length equal to 1,650,763.37 wavelengths of the orange light emitted by the Krypton atom of mass 86 *in vacuo*! However, to shift from a measurement system that does afford concrete meaning to one that is more abstract needs careful support, and has historically taken time and much effort, as seen in different country contexts.

Historically, the earliest measures still in use are 'anthropomorphic', corresponding to parts of the human body; such as, the foot (say, to mark distances while sowing a crop), the pace, or the elbow or 'ell' (to measure cloth, etc.). The width of the loom for textiles, the size of the milling equipment for glass 'panes', or the size of the 'pig' for raw iron have given rise to units shaped by the production process. Similarly, transportation has defined the cartload, the basketful for particular crops, and the boatload for sand, soil, etc. Discussing these with adults enhances their understanding (and ours) of the context of various systems of measurement and facilitates their ability to adopt other standard ones. There is a striking dominance of qualitative 'value' over purely quantitative considerations in the social thinking of pre-industrial societies. For instance, traditional land measures were not directly 'addable', and accounted for the quality of soil, labour-time for tilling, or the amount of seed needed. Similarly, the 'measure sold' was different in size from the 'measure bought'; a heaped container is bought from the farmer while a flat one is sold to a customer at the same price, to take care of the costs of transaction; even today we are given 13 glass bangles at the rate of a dozen (to cover possible breakage). Such forms of measurement co-exist in everyday practices with the standard systems, but most conventional adult educators refuse to appreciate their ingenuity and significance, dismissing those as 'crude' or 'primitive', thus alienating their learners.

While developing the numeracy materials, we tried to discuss with the volunteers how historically the metric system had sought a shift from the synthetic-qualitative manner of thinking about a measure to a more abstract-quantitative process. Through the use of the metric system, we normally tend to abstract one out of the many different qualities of diverse objects, say, the length, and view them all from this single perspective. The pace of a woman, the length of her sari, the height of a tree, the thickness of a sheet of paper, or the diameter of the earth are all collapsed to the same measure of length – the 'metre'. Further, the divisibility and cumulativeness of the metric system enables us to compare magnitudes diversely different – from that of the atom to the cosmos – by just sliding up through orders of ten. This however is not possible with the conventional measures.

Teaching through a plethora of creative activities to estimate and measure had led us to the popular concept of a village numeracy fair – the Metric *Mela* – conducted by the volunteers and learners themselves, where the entire village turned out in all its finery to participate. The following excerpt from the Numeracy Handbook (Rampal, Ramanujam and Saraswathi 1998 48–50) describes such a numeracy fair or *mela*, how it was conducted and was named so, through the creative agency of the adult learners:

> Even as you approach the *mela*, you can hear the songs on the loudspeaker. You wish somebody would reduce the volume, but it is undeniable that the din caused does add to the festive atmosphere. Indeed it seems to go with the festoons, streamers and the general riot of colour, with noisy children running about adding to the *mela* mood.
>
> A couple of volunteers come forward to welcome you, assuring that you are about to have a 'totally new' experience, and that this *mela* is entirely run by their students, the neo-literates of the village. You have heard this before, when they went door to door yesterday inviting everyone in the village to come, and even offered attractive prizes! Even as you join the queue, you ask why 'metric' *mela,* but receive no clear response. The volunteers, young girls themselves, giggle a bit and are mysterious – you'll find out soon. ... Even as you move along, you are intrigued by your red card [see Figure 11.1], and you take a look at it. It has a big table with each row having a description and some blank entries. Passing over routine items like height, weight etc. you are intrigued by entries like 'weight of a feather?', 'length of *lauki* (gourd)' etc. Wait a minute, what is that – 'length of nose'?! They are not going to measure the length of your nose, are they!?

Name : Male/Female	Card No. Address :
My height : cms	My weight: kgs
Estimate	
Length of the *lauki* (gourd):	Length of my nose:
Length of a chalk :	Length of my little finger:
Weight of an egg :	Weight of the cabbage:
Weight of a feather:	
Which has more water?	
Volume of water in the bucket:	
Volume of water in the bottle:	

FIGURE 11.1   Your 'metric' card – how good is your estimate?

The first stall you go to has a person with a measuring tape, who measures your height, enters it in your card and makes an entry in her own register as well. In the next one, as expected, someone measures the length of your nose! While you now have a pretty good idea of what the *mela* is about, and why it is a metric *mela*, the whole thing falls into place only in the evening, when there is a festive cultural programme followed by the much-awaited prize-giving ceremony. This is a virtual riot as there are prizes in a most interesting variety. There is one for the person with the longest nose, and one for the person with the shortest nose. The person who got the closest estimate for the weight of cabbage gets the cabbage itself as a prize, and similarly the one who got the length of the drumstick right gets the drumstick. Prizes for the tallest, shortest, heaviest, lightest, … Indeed there are prizes for almost everyone, and you get the prize for … getting the weight of the chicken feather correctly!

The prizes are given away by the neo-literate who ran the stall, and this is in itself a novel experience for her and for the village. She also talks about how many got close to the answer, how many gave wild guesses (with some examples, causing much mirth). Interestingly, in all the stalls, though people gave their answer in whatever units they pleased, they have been converted to metric units before recording. Thus the term 'metric *mela*!'

National primary textbooks: mediating social practices

The NCF 2005 position paper for the teaching of mathematics (NCERT 2006a) calls for a multiplicity of approaches, for liberating school mathematics from the tyranny of the one 'right' canonical answer, through learning environments that promote mathematisation, invite participation, and offer *every* child a sense of success. A 'participationist' (Sfard 2008) vision of learning mathematics, unlike the acquisitionist approach, requires that learners begin by participating in collective mathematical discourses, and progressively learn to communicate mathematically with themselves.

While developing the new primary mathematics textbooks for NCERT (2006b, 2007, 2008), I drew upon the Numeracy Handbook and our experience of the Literacy Campaign, to infuse people's social practices of everyday mathematics – of dealing with numbers, measurements, estimations, shapes, symmetries and aesthetics – into the primary school textbooks. Moving beyond the rigid boundaries of the traditional school subject, we included lived contextual examples from art, craft, architecture and music. We also carefully sought narratives of real-life protagonists from across the country to inspire and animate the thematic mathematics chapters. The textbook was designed to resonate with children's 'lived' resources (Gueudet, Pepin and Trouche 2012), through diverse genres of expressive narratives, folklore, (auto)biographical stories, household recipes, travelogues, diaries, letters, records such as birth certificates, humour, fantasy, etc. We moved away from traditional inanimate illustrations which offer stunted, stereotyped and monotonous images, to diverse representations including folk and tribal art, photographs, children's art, cartoons, and contemporary art informed by multicultural sensibilities. Working closely with chosen artists, we attempted to design each page as a visual text, which could be processed by children in a non-linear manner. Though separate teachers' handbooks could not be developed, the teachers' notes shared reasons for dealing with concepts differently, observations about students' thinking, and suggestions for out-of-class activities in specific cultural contexts, all of which were aimed at persuading teachers to promote participation in place of the dominant modes of transmission.

The National Curriculum Framework 2005 had recommended breaking down the rigid boundaries between different subjects to help develop a more holistic approach of learning from the child's environment and culture. This was attempted in the NCERT primary textbooks within the three learning areas – language, math and environmental studies. Thus, for instance, we took the theme of 'mapping', which is usually dealt with inadequately in the social studies and geography books, to cut across all the three primary school areas. Chapters for the mathematics textbooks were developed around the concepts of projections, plans and perspective, aerial views of a site or an object from different heights (as seen by a mouse riding on a hydrogen balloon), directions, scaling, representation, etc. Iconic and pictorial maps were made for specific contexts, linked to narratives, such as children finding their way to the beautiful monument of Taj Mahal, which gradually progressed to abstract schematic representations. In addition, several creative formats were used, such as diaries or travelogues for a historical monument, visuals and pictorial maps, as well a treasure-hunt game. Simultaneously, the environmental studies textbook developed the ideas of mapping through different chapters, on a historical fort, and one on 'Sunita Williams in Space', based on the true experiences of a NASA astronaut of Indian origin, who eloquently described her thoughts as she looked at the earth from space, and even poignantly wondered where the 'lines' or boundaries got drawn which, from there, could not be seen between India and its neighbouring countries.

While most of the chapters were based on specific concepts, we developed some thematic chapters to deal with issues of work, entrepreneurship, heritage, craft knowledge, history of monuments and pre-historic cave paintings, etc., using contexts that invoked and integrated concepts already learnt. For instance, 'Building with Bricks' in Class IV (NCERT 2007) begins with the true instance of a school being built by local masons, Jamaal, Kaalu and Piyaar, who visit the nearby mosque to observe the amazing variety of floor patterns built by their ancestors three hundred years before. They return inspired and make their own brick designs for the school courtyard. With modern bricks different from the older thinner ones, the masons generate different symmetries and patterns, which students are encouraged to analyse in photographs we had taken during our visit. The chapter goes on to measure a brick, to study its faces, see its projections and how (as the first example of a cuboid) it can be represented in two dimensions. It prompts students to observe photographs of other brick patterns from different parts of the country and analyse the designs in traditional architecture. There are examples from the work of Laurie Baker (without naming him), a Gandhian architect who devised low-cost environment-friendly buildings. It finally travels to a brick kiln, to understand the process through visuals; students are first introduced to the large number 'one hundred thousand' (a lakh), as the number of brick kilns in the country.

The process of thinking of large numbers (as had been done in the adult numeracy curriculum), relatively and in familiar contexts, through orders of magnitude, is adopted throughout the books. For example, they connect 100 with the scoring of a century by a famous cricketer, or are asked to recall where they have heard of a 'lakh' (one hundred thousand). Similarly, the number one crore (ten million) is first introduced in the thematic chapter 'The Fish Tale', as 'the number of people whose lives are related to fish – who catch fish, clean and sell them, make and repair nets, and boats, etc.' (NCERT 2008, 10). Indeed, 'asking students questions, such as, "How long does it take to count to 1,000?" or "Have you lived more or less than 1,000 days?" provides them an opportunity to think about 1,000 in a personal context, thus helping them better understand the size of 1,000 in a variety of contexts' (McIntosh, Reys and Reys 1992, 16).

'The Junk Seller' is based on the true and unusual story of a young woman Kiran, who had, against all odds of living in a poor, highly patriarchal society, managed to set up her own enterprise in the city of Patna. She narrates her struggle, her early dislike of math in school and her acknowledgement of how it is now an integral part of her present vocation, which has indeed helped change her life and the situation of her family.

When I was young my father died in an accident. So my mother worked as a servant in some houses. We had a difficult time. I had to leave school after Class VIII. I wanted to study more but my mother got me married. My husband's family lived in a mud house. … He had a tea stall. I thought of starting my own business. I thought I should open a bangle shop or a tailor shop. But my uncle said that we could earn a lot by opening a junk shop. People laughed and teased us about our work. They called it *ganda kaam* or 'dirty business'. But I did not think so. I knew this idea would work.

(NCERT 2007, 60–61)

Through a visual narrative with on-site photographs, the unit deals with her loans, her junk sorting and selling, hiring of collectors, recycling of materials, etc. It challenges several prevailing notions of gender and mathematics, the stigma of 'dirty work' attributed to certain castes and their supposed low position in society, and the traditional focus on a 'great man' as a role model. It inspires young women with a sense of 'social agency' to develop their entrepreneurial abilities to transform lives. A year later when we met Kiran, she proudly told us that many visitors come to her shop only to see the 'textbook hero'.

Interestingly, this focus on cultural relevance and real life contexts caught the public imagination. Leading national newspapers and TV channels followed the development of the new textbooks. Young reporters met us to understand the changes and recalled their own unpleasant school experiences. Full page or lead stories, normally unusual to see on this theme, came with headings such as 'NCERT's Bold New Experiment Brings Maths Closer to Life'. This story began with: 'Ever thought you could study geometry from brick patterns on the walls of a tomb in Murshidabad? Or arithmetic from a junk-seller in Patna? Well that's what the new Class IV math textbook by NCERT is all about: maths and real life' (Mukherji 2007, 4).

In general, state textbooks do not reflect culturally responsive perspectives. Conservative curricula allow a tokenistic approach where diversity may at best be represented in a celebratory manner, without critical engagement about issues of difference, discrimination and dominance. We had noted that chapters on the concept of 'time' in textbooks across countries routinely deal with clocks and calendars, describing various units and devices of measuring time, and 'informing' the child that a day is made up of 24 hours, an hour of 60 minutes, while a week is seven days, and so on. Why so? Indeed, we found that a richer canvas could be drawn when learning about 'time' through different cultural contexts, including their subjective notions of 'experienced time', while gradually scaffolding their understanding through progressively more challenging tasks and investigations. The Class III chapter 'Time Goes On' begins with asking children to correct some funnily jumbled up time markers used in a visual story, depicted in a folk art *pattachitra* style (which traditionally uses cyclic representations for time). Drawing upon children's common sense and real-life observations and estimations of, for instance, how long it takes for a fruit to fall from a tree, for curd to set, for a litre of milk to boil, or a baby to come out of her mother's stomach (a deliberate attempt at subversive humour), it leads them to think of several processes that elapse in different orders of time – in years, months, hours, minutes or seconds. We ask:

> Have you seen someone knitting a sweater? Or, someone weaving a cloth? Do try to find out from a potter how long it takes to make a pot. Also tell us if you take hours or minutes to have your bath! (Is it years since you last had one? Ha, Ha!)
>
> *(NCERT 2006b, 96)*

Our voice tries not to be moralising, but empathetic, to help value the person, her craft and labour. Cultural narratives are used, such as about celebrating 'a thousand full moons' as is still done in some places when a person completes eighty years. The chapter exposes children to the fact that 'experienced time' can be starkly different – even as it asks urban learners to compare their daily routine, from morning to night, with that of a village girl. It also uses the concept of a number line for the first time, representing the real 'timeline' of a woman's life, who narrates when time flew, as she busily raised her young children, and when it stood still – at the time her husband took ill and died. This woman had been a participant of the literacy campaign, and had described to us how she was helped by the literacy volunteers to legally fight for her rights from her husband's scheming relatives, who declared her a 'witch' to snatch her land. There was much deliberation in the textbook team before we incorporated this particular exercise, as to whether such social realities could find a place in the primary classroom. In fact, when some urban middle-class teachers wondered if 'witches' should be discussed with young children, it was discussed why no questions were generally asked about the suitability of a Harry Potter book for young children or even about viewing violent cartoons of witches on television; then why was the mention of a real person falsely declared a 'witch' so problematic? Instead, wasn't it important and inspiring for children, rich and poor, to know about the agency of the woman and others who had attempted to resist this exploitative practice? The decision was thus taken to consciously depart from the approach of curricular 'infantilisation' or 'Walt Disneyfication' (Giroux 1994). The conventional approach, in fact, misguidedly resorts to comforting cartoon characters and believes that children must be 'protected' from the harsh realities and injustices of the real world – at least in textbooks, no matter what actually happens to them in the real world (Rampal 2015). In this way, it underestimates and also hampers children's critical agency in being able to deal with social realities. It also presents an unreal homogenised 'universal' world of comic cartoon figures, devoid of real flesh and blood characters that can inspire children.

Symbolic mediation between the home and school language is an important dimension of creating possibilities for meaning making. However, in a misguided move mathematics (along with science) is increasingly being taught in English in many schools across the country, though the regional language is chosen for the social sciences. Working on the NCERT primary textbooks (produced in Hindi, English and Urdu and given to the states to adapt and translate into other languages), we consciously forged an 'articulation' (Skovsmose 2012) of different discourses through multiple genres. We gave voice to children's imagery and intuitive, tentative ways of thinking and also did not concede to the premature use of technical terms where more accessible alternatives were available.

Transacting a mathematics curriculum in a sociocultural framework 'to read the world' also requires what Freire (1970, 62) calls 'problem-posing pedagogies', as distinct from problem solving ones, so that education 'involves a constant unveiling of reality ... that strives for *critical intervention* in reality'. It requires distinguishing between using mathematics in real world settings, usually limited to shopping, travelling, or building, from asking students to critically investigate issues of injustice, through a sense of collective social agency (Gutstein 2006).

A work in progress

The syllabi and textbooks are still used by all schools under the Central Board of Secondary Education, though states have their own State Councils and either adapt and translate the NCERT books or produce their own textbooks. However, much of the challenge of orienting teachers to change pedagogical interactions within the classrooms still remains. The present dispensation is now working on its own plan for reform, which has sent some worrying signals. Given the resource starvation of most of our elementary schools, government textbooks developed at the national level by NCERT or at the state level by the SCERTs (provided free of charge) form the only curriculum materials available for the majority of children and teachers, while private schools use more expensive books (of often poor quality) published by private publishers. There are mostly no concrete objects, games or manipulatives in the classroom, no school library, and teachers' education rarely steers them to mediate such resources. Low-paid contractual appointments, frequent testing, and a concerted corporate push towards privatisation further demoralises teachers, especially those working in schools of the poor. Curricular reform involves sustained, unhurried and layered negotiation, within and outside the system, from policy documents to classroom practices and examinations, involving administrators, teachers, parents, teacher educators and the media, to change mindsets about how children learn, how that may be assessed, and what basic provisions are conducive for that learning to happen. It has happened, in some places at some times, but is nevertheless an ongoing struggle.

References

Addey, C. (2017) "Golden relics and historical standards: How the OECD is expanding global education governance through PISA for Development" *Critical Studies in Education*, 58(3): 1–15

Barton, D. and Hamilton, M. (eds) (2012) *Local Literacies: Reading and Writing in One Community* London: Routledge

Boaler, J. (2008) *What's Math Got to Do with It? Helping Children Learn to Love Their Most Hated Subject--and Why it's Important for America* New York: Viking

Civil, M. (2007) "Building on community knowledge: An avenue to equity in mathematics education". In Nasir, N. and Cobb, P. (eds), *Improving Access to Mathematics: Diversity and Equity in the Classroom* New York: Teachers College Press

Cobb, P. and Bowers, J. (1999) "Cognitive and situated learning perspectives in theory and practice" *Educational Researcher*, 28(2): 4–15

Daniels, H. (2001) *Vygotsky and Pedagogy* London: Routledge Falmer

Freire, P. (1970) *Pedagogy of the Oppressed* New York: Herder and Herder

Gee, J.P. (2004) *Situated Language and Learning: A Critique of Traditional Schooling* New York: Routledge

Gerdes, P. (1985) "Conditions and strategies for emancipatory mathematics education in underdeveloped countries" *For the Learning of Mathematics*, 5(1): 15–20

Ghose, M. (2007) *Exploring the Everyday: Ethnographic Approaches to Literacy and Numeracy* New Delhi: Nirantar

Giroux, H.A. (1994) *Disturbing Pleasures: Learning Popular Culture* London: Routledge

Government of India (2009) The Right of Children to Free and Compulsory Education Act. New Delhi: Ministry of Law and Justice

Greer, B., Mukopadyay, S., Powell, A. B., and Nelson-Barber, S. (2009) *Culturally Responsive Mathematics Education* London: Routledge

Guendet, G., Pepin, B., and Trouche, L. (2012) *From Text to 'Lived' Resources: Mathematics Curriculum Materials and Teacher Development* New York: Springer

Gutstein, E. (2006) *Reading and Writing the World with Mathematics: Toward a Pedagogy for Social Justice* New York: Routledge

Hamilton, M. (2016) "Imagining literacy: A sociomaterial approach". In Yasukawa, K. and Black, S. (eds), *Beyond Economic Interests: Critical Perspectives on Adult Literacy and Numeracy in a Globalised World* Rotterdam: Sense

Jordan, C. (2012) "Running the numbers: A conversation". In Mukhopadhyay, S. and Roth, W. M. (eds), *Alternative Forms of Knowing (in) Mathematics: Celebrations of Diversity in Mathematical Practices* Rotterdam: Sense Publishers

Kula, W. (1986) *Measures and Men* Princeton, NJ: Princeton University Press

Lave, J. (1996) "The practice of learning". In Chaiklin, S. and Lave, J. (eds), *Understanding Practice: Perspectives on Activity and Context* Cambridge: Cambridge University Press

Lave, J. and Wenger, E. (1991) *Situated Learning: Legitimate Peripheral Participation* Cambridge: Cambridge University Press

McDermott, R. P. (1996) "The acquisition of child by a learning disability". In Chaiklin, S. and Lave, J. (eds), *Understanding Practice: Perspectives on Activity and Context* Cambridge: Cambridge University Press

McIntosh, A., Reys, B.J. and Reys, R.E. (1995) "A proposed framework for examining basic number sense". In Murphy, P., Selinger, M., Bourne, J. and Briggs, M. (eds), *Subject Learning in the Primary School: Issues in English, Science and Mathematics* London: Routledge and Open University

Mukherji, A. (2007) "NCERT's bold new experiment brings maths closer to life" *Times of India*, March 24, p. 4

Nasir, N.S. and Cobb, P. (2007) *Improving Access to Mathematics: Diversity and Equity in the Classroom* New York: Teachers College Press

NCERT (National Council for Educational Research and Training) (2005) *National Curriculum Framework* . Available at http://www.ncert.nic.in/html/pdf/schoolcurriculum/framework05/nf2005.pdf (accessed 18 January 2018).

NCERT (National Council for Educational Research and Training) (2006a) "Position Paper 1.2 of the National Focus Group on the Teaching of Mathematics". Available at http://www.ncert.nic.in/new_ncert/ncert/rightside/links/pdf/focus_group/math.pdf

NCERT (National Council for Educational Research and Training) (2006b). *Math-Magic: Textbook for Mathematics for Class III*. New Delhi: NCERT. Available at http://ncert.nic.in/textbook/textbook.htm?cemh1=0-14 (accessed 16 February 2018).

NCERT (National Council for Educational Research and Training) (2007). *Math Magic: Textbook for Mathematics for Class IV*. New Delhi: NCERT. Available at http://ncert.nic.in/textbook/textbook.htm?cemh1=0-14 (accessed 16 February 2018)

NCERT (National Council for Educational Research and Training) (2008). *Math Magic: Textbook for Mathematics for Class V*. New Delhi: NCERT. Available at http://ncert.nic.in/textbook/textbook.htm?cemh1=0-14 (accessed 16 February 2018)

Nunes, T., Schliemann, A.D. and Carraher, D.W. (1993) *Street Mathematics and School Mathematics* Cambridge: Cambridge University Press

Rampal, A. (1992) "A possible 'orality' for science?" *Interchange*, 23(3): 227–244

Rampal, A. (2003) "Counting on everyday mathematics". In Saraswathi, T. S. (ed), *Cross-cultural Perspectives in Human Development: Theory, Research, and Applications* New Delhi: Sage Publications

Rampal, A. (2013) "Lessons on food and hunger: Pedagogy of empathy for democracy" *Economic and Political Weekly*, XLVIII(28): 50–57

Rampal, A. (2015) "Curriculum and critical agency: Mediating everyday mathematics". In Mukhopadhyay, S. and Greer, B. (eds), *Proceedings of the 8th International Mathematics Education and Society Conference* Portland, OR: Ooligan Press

Rampal, A., Ramanujam, R. and Saraswathi, L.S. (1998) *Numeracy Counts!* Mussoorie: National Literacy Resource Centre, LBS National Academy of Administration. Available at https://archive.org/details/NumeracyCounts (accessed 18 January 2018).

Rampal, A., Ramanujam, R. and Saraswathi, L.S. (2000) *Zindagi Ka Hisaab* Mussoorie: National Literacy Resource Centre, LBS National Academy of Administration. Available at https://archive.org/details/ZindagiKaHisab-Hindi-MathsFromDailyLife (accessed 18 January 2018).

Rogoff, B. (1995) "Sociocultural activity in three planes". In Wertsch, J. V., del Rio, P. and Alvarez, A. (eds), *Sociocultural Studies of Mind* Cambridge: Cambridge University Press

Sangh, H. C. (All-India Education Board) (1938) *Report of the Zakir Hussain Committee on Basic Education* Maharashtra: Wardha

Sfard, A. (2008) "Participationist discourse on mathematics learning". In Murphy, P. and Hall, K. (eds), *Learning and Practice: Agency and Identities* London: Sage, London

Skovsmose, O. (2012) "Towards a critical mathematics research programme?". In Skovsmose, O. and Greer, B.(eds), *Opening the Cage: Critique and Politics of Mathematics Education* Rotterdam: Sense Publishers

UNESCO (1990) *World Declaration on Education for All*. Available at http://unesdoc.unesco.org/images/0012/001275/127583e.pdf (accessed 18 January 2018).

Vinoba (2014) *Thoughts on Education* 8th edn, (translation by Marjorie Sykes) Varanasi: Sarva Seva Sangh Prakashan

Yasukawa, K and Black, S. (eds) (2016) *Beyond Economic Interests: Critical Perspectives on Adult Literacy and Numeracy in a Globalised World* Rotterdam: Sense Publishers

12

CRITICAL HUMANISTIC PEDAGOGY IN THE CONTEXT OF ADULT BASIC EDUCATION

Making sense of numeracy as social empowerment

Rebecca Nthogo Lekoko, Shanah Mompoloki Suping and Obusitswe Pitso

Introduction

This chapter proposes an approach to teaching adult basic numeracy grounded in a view of numeracy as a form of mathematical skills best understood and applied in specific cultural contexts. Culture here is understood as people's ways of life, for instance, how certain groups of people solve social problems, and the language they use for carrying out day-to-day life activities. Positing that numeracy activities are grounded in cultural practices means that people have established ways of how to innovate, think logically and solve problems using numbers, images and other mathematical codes (D'Ambrosio 1997). Based on this understanding of numeracy as knowledge that can be used to address real-life challenges, the term social empowerment is introduced to capture these life-enhancing properties of numeracy. Social empowerment implies that those who learn numeracy can use it to improve their lives and those of other members in their community.

The concept of social empowerment is consistent with the idea of ethnomathematics. Milton Rosa and Daniel Orey (2011) present ethnomathematics as mathematical concepts that are related to the cultural and daily experiences of people studying mathematics. To call for the use of ethnomathematical approaches is to urge educators to make learning of mathematics more relevant and meaningful to learners by helping learners to make meaningful connections between what is taught and what they already know, hence deepening learners' understanding. As Rosa and Orey (2011, 1) further state, using an ethnomathematical approach is good in helping learners to develop intellectual, social, emotional and political acuity using unique cultural referents, and this greatly helps 'them maintain their identities while succeeding academically'.

The brief properties of ethnomathematics presented in the preceding paragraph embed characteristics of a critical humanistic philosophy that this chapter presents as a theoretical framework. The ultimate goal of a critical humanistic perspective is to see learners apply their learning in real-life contexts. Those who profess a humanistic perspective see learning as a tool to develop well-rounded, critically conscious individuals with responsibility to use knowledge gained from the learning environment to address circumstances of their lives (Aloni 1999). Both ethnomathematics and humanistic perspectives consider educational issues devoid of cultural, social, and economic context a serious ignorance of what learning is all about (Aloni 1999; Millroy 1992). The ideas of ethnomathematics and critical humanistic perspectives blend well through their emphasis on developing learners who are able to apply what they learn in real life situations. This chapter calls for the development of ethnomathematically literate adults, that is, adults who are empowered to use numeracy skills to address everyday life needs. Thus, when ethnomathematics and critical humanistic perspectives are woven together, they shape a pedagogy that presents the use of familiar contexts and lived experiences in the teaching of adult basic numeracy as apt cultural practices to link what is learned to people's life needs. The suggested pedagogy is anchored to the current practices of adult basic education within the Botswana National Literacy Programme (BNLP).

The BNLP targets Batswana youth and adults aged 15–45 years old 'who are not attending formal school; who have never attended school or who have attained standard 4 or below or its equivalence' (Statistic Botswana 2014, 2). This population is the target of the proposed pedagogy for a socially empowering numeracy. Empowerment can be effectively promoted in situations where learning activities draw from or use familiar contexts and lived experiences. This type of learning builds learners' self-confidence which makes it possible for facilitators to take learners through a learning journey which, while exploring basic numeracy, can also engage learners more deeply with critical forms of analysis of mathematics/numeracy. The idea of social empowerment is brought in to underscore the humanistic philosophy that presents learning as a tree of knowledge that should nourish a tree of life (Aloni 1999). As Paul Ernest (2010, 23) contends, the goal is to develop a socially-numerate citizen who can 'critically understand the uses of mathematics in society: to identify, interpret, evaluate and critique the mathematics embedded in social, commercial and political systems'.

To call for ethnomathematical and humanistic approaches to teaching adult basic numeracy acknowledges some shortcomings within the character of the BNLP. It is befitting to straightforwardly note that some progress has been made in the BNLP to orient and shift practices from teaching that stressed memorisation at the expense of functional and productive skills (Maruatona and Mokgosi 2006), though more effort is still needed. Maruatona and Mokgosi recommended a curriculum that draws from problems and issues of communities and the nation at large. Their work is in line with this chapter's recommendation of a pedagogy enriched with humanistic and ethnomathematical perspectives so as to make BNLP more responsive to the real life needs of its beneficiaries.

This chapter starts with a general historical background of teaching and using mathematical ideas in everyday cultural settings. As the chapter unfolds, key aspects of numeracy for life and culturally-based learning are presented to underscore the basic, but often ignored, fact that every meaningful learning activity is useful in real-life contexts. This idea gets impetus from the great African philosopher, Julius Nyerere (1973, 141), who once declared that 'to live is to learn; and to learn is to try to live better'. This chapter concludes by suggesting a framework of teaching/learning numeracy skills using a life case-study approach to empower learners to apply skills learned in their daily living activities.

Background

Numeracy is recognised as an essential form of basic mathematical skills needed for use in all aspects of life. It would be impossible to live a normal life now or in the future without making use of mathematics of some kind (Cockcroft 1982 cited in Glevey 2009). This makes it imperative that the teaching of numeracy be done cautiously to ensure 'a better fit between what it takes to live in today's societies' and what learners learn (Davis 1993, 10). The approach uses cases of numeracy to engage learners in social processes such as reciprocal dialogue that would lead to the development of relevant skills and attitudinal attributes necessary for adults to appreciate and use numeracy knowledge as an indispensable aspect of their lives.

A case approach is preferred for its potential to promote practical numeracy skills needed for today's modern economies. A case approach implies the use of real-life experiences. As cited in Lekoko (2015), LawNerds.com (n.d.) defines cases as 'the stuff of real life'. In this chapter, the authors share real-life experiences of the use of ethnomathematics drawn especially from the African context like Botswana. Cases used have been selected for their practical value and accuracy in helping readers understand numeracy as social empowerment.

To be socially empowered implies the ability to use mathematics to address both personal challenges and those that impact wider social life, both in the present and in the future (Malloy 2008). Furthermore, a person who is socially empowered possesses practical knowledge and self-confidence, acting individually and collectively to address real-life challenges. This perspective becomes more pronounced in the context of numeracy taught within adult basic education (ABE), since ABE participants specifically learn to acquire functional skills to use in their respective life circumstances. The concept of ethnomathematics is usually used to explain this need for acquisition of skills that can be applied to real-life circumstances.

From the authors' perspective, ethnomathematics has strong roots in the African traditional or indigenous ways of learning. African traditional artefacts, for example, beadwork, carpentry, weaving, patchwork, curving, designing, and decorating would not have been possible without some mastery of numerical literacy. Numeracy of this nature has been learned within an action-oriented culture in which learning was not separated from the rest of people's lived activities, compared to some form of decontextualised formal learning of mathematics in schools. Based on the need for this contextual savvy, an approach for teaching numeracy skills using familiar contexts is recommended, specifically using the technique of moving 'from known to the unknown'.

Teaching from known to the unknown is a concept that presents learning as a process that does not begin in formal institutions; people always learn informally in their respective environments. The BNLP recommends that these prior experiences be acknowledged. Adults who attend the BNLP bring rich experiences or pre-existing knowledge and skills in the learning environments. This prior knowledge becomes helpful when they are taught mathematical concepts related to such prior understanding. Prior experiences can actually help them to remember or think about how they have learned the concept in the past and applied the information gained in their life circumstances; in this manner, they use prior experience to build on or create new knowledge being taught in the classroom atmosphere.

To a certain extent, the BNLP still follows the traditional approach of teacher-centred approach, and as Mpofu and Youngman (2001) have stated, this approach has been relatively ineffective in improving adult literacy levels. Calling for teaching from known to the unknown is to recommend a shift from teacher-centred to learner-centred approach and its accompanying practical literacy activities. Understood from the perspective of what Rogers (1999 cited in Mpofu and Youngman 2001, 4) calls 'real literacies', it is assumed 'that most adults already engage in some literacy tasks in their daily lives (like reading the destination on buses or counting their change)' and these prior experiences are the basis for recommending an approach that takes learners through processes that recognise that they already know something about what they are taught. In this way, the prior experience is credited as a stimulus for current learning.

Adults, as globally known, desire to relate new information to their experiences and if new learning does not fit in with what adults already know, they may reject it (Knowles 1984; Birkenholz 1999). Thus, moving from the known to the unknown is a strategic approach to relate past to present learning. This type of approach has many benefits. Like Kane (2002) contends, as a practical approach to learning, it promotes the interactive learner-centred approach and solution-centred activities. It further gives learners confidence to learn and affirm their cultural identity in the process of learning (Schlöglmann 2006). It reflects what is known about adult learning, that adults' 'prior life experiences play a key role in the learning activities, they rely on these experiences as a resource from which they can learn new things' (Pedsen 2013, 19). The discussion here does not assume a causal relationship but presents a high probability of successful learning if the known to the unknown approach is followed.

Historical overview of learning of mathematics in the African context

Mathematical ideas exist in all cultures, and, as Ascher (1991) says, the way mathematical literacy is experienced, expressed, acquired, and used can vary from culture to culture. This perspective, however, is not oblivious to traditional conceptions that present mathematics as a culture-free and value-free phenomenon (Ellerton and Clements 1990). The focus and scope of this chapter does not permit dwelling on these divergent perspectives. Rather, this chapter argues for an anthropological awareness that different cultures can produce and use different forms of mathematics rather than the Eurocentric part cularity or universality of mathematics (D¢Ambrosio 1985). Babungura (2002) calls attention to these unique contexts, persuading people to study mathematics from various cultural perspectives. According to Ascher (1991, 1), doing this would reveal 'a global, multicultural view of mathematics'.

Ethnomathematics emanates from a viewpoint that mathematics as a cultural product is created and belongs to people (Powell and Frankenstein 1997). Accordingly, in trying to understand the everyday use of mathematical literacy, D'Ambrosio's (1997) idea of performativity (performative context) is plausible. He argues that mathematical ideas are continually generated and used to respond to life issues in specific contexts. D'Ambrosio further explains that instances of these codes, norms, rules, and values are instruments of analyses and explanations of actions. Examples of these codes, norms, rules, and values include cardinality and ordinality, counting and measuring, and sorting and comparing, and these take different forms according to the cultures in which they are generated, organised and accepted. What is accepted as meaningful and useful is determined by the culture to which one belongs; for example, for Africans who lead an agriculturalist way of life skills such as making a yoke and span, or a sledge for cattle are examples of meaningful ways of engaging with numeracy. Additionally, artefacts, such as carved and painted masks associated with religious practices, or the decoration of common household objects such as utensils and baskets link with ethnomathematics. Figure 12.1 illustrates some of these traditional artefacts. The skills used to create these artefacts are not a result of having learned through formal schooling, but come naturally as people interact or respond to life responsibilities and challenges in a given cultural setting.

FIGURE 12.1 African traditional artefacts illustrative of mathematical use (source: personal photography and artefacts)

Another interesting example of mathematics use primarily driven by African cultural practices is the concept of time. Mbiti (1988, 20), for example, contends that in the traditional African culture, time is understood according to significant events, for example, 'a day is understood in terms of the events'. By illustration, in Botswana, the time one wakes up in the morning used to be based on the crow of a cock and not on numeric time, like 6:00 a.m., which is a specific Western idea of time. Furthermore, months of the year attain their names from events associated with them (Mbiti 1988). The month of June in Botswana is called 'Seetebosigo' (visiting at night is not advisable) meaning 'don't visit at night'. Since Botswana is a semi-arid desert climate, it can have extreme hot and cold months. At the heart of winter is the month of June, which can be very cold hence the name Seetebosigo. Even though most people in Botswana have never experienced snow-cold weather or below freezing temperature, people are warned of June's cold nights and advised to remain indoors and under covers. Events or actions matter more than when (numerical value) they happened hence Mbiti's assertion that, 'it does not matter whether the hunting month lasts 25 or 35 days, the event of hunting is what matters much more than the mathematical length of the month' (Mbiti 1988, 20). As the given examples demonstrate, most people learn in accordance with the dictates of their different cultures (Sheffield 1972). This way of learning made Africans independent producers of knowledge for their survival thus affirming that knowledge is produced locally and used primarily for survival.

Humanistic perspectives in Botswana's National Literacy Programme

BNLP is the government's initiative to provide basic education and literacy for adults. Among its objectives is to eradicate illiteracy and enable the participants to apply knowledge in developing their cultural, social, and economic life (Maruatona and Mokgosi 2006). Adult basic literacy programmes have been oriented towards this humanistic perspective after realising that the past curricula were fundamentally traditional and stressed memorisation at the expense of functional and productive skills (Gaborone et al. 1987 cited in Maruatona and Mokgosi 2006). Greater focus has now shifted away from academic skills of reading, writing, and numeracy to a curriculum that draws from problems and issues of communities, societies, or the nation at large. According to Maruatona and Mokgosi (2006), the definition of literacy has changed from the original simplistic concept of ability to read and write with understanding and is now aligned with the cultural functional skills and thus viewed as

> a responsive and context specific multidimensional lifelong learning process designed to equip beneficiaries with specialised knowledge, skills, attitudes and techniques to independently engage in practices and genres involving listening, speaking, reading, writing, numeracy, technological functioning and critical thinking required in real life.
>
> *(Maruatona and Mokgosi 2006, 24)*

UNESCO (2011) observed that the BNLP has an ambitious goal. As the largest government-sponsored programme for non-formal education, its launch in 1981 was rationalised as part of the great social reform. In its wider sense, the NLP was seen as a part of strategies to

> eradicate historical socio-economic inequalities and high adult illiteracy and create a cohort of educated people with skills to meet the demands of a developing, rapidly changing society and economy as well as to empower previously disadvantaged and marginalised communities in order to enable them to be self-reliant and to improve their standard of life.
>
> *(UNESCO 2011, para. 1)*

Adopting this comprehensive stance meant implementing a diverse and much broader curriculum including Basic Literacy, the Literacy at the Workplace Project, Income Generating Projects, the Village Reading Rooms Project, and English as a Second Language (UNESCO 2011). Logically, as presented here, NLP in Botswana signifies the value of the humanistic philosophy that promises to open up possibilities of developing human dignity and intellectual freedom towards the best and highest capability of using knowledge to solve real-life challenges (Aloni 1999). However, as Maruatona and Mokgosi (2006) state, the main challenge remains teaching strategies, as education in Botswana is still marred by mathematics teaching strategies based on teacher-directed learning mainly constituted by memorisation and recitation. This abstract memorisation of mathematical concepts gives little room for the development of life skills. Critics like Epstein and Yuthas (2012) believe that imposing 'bookish' and foreign ways of learning mathematics is a way to undermine the ability of the local people. This chapter thus proposes a pedagogical framework to engage adult learners using familiar cases as the starting point and moving towards the development of problem-solving abilities.

A proposed approach to teaching basic numeracy to non-formal adult learners

The proposed approach for teaching adult basic numeracy relies primarily on constructing mathematical knowledge from the known to the unknown and using social learning processes like dialogue. The aim is to expose or unfreeze mathematical ideas embedded in cultural practices and hence illustrate learning of mathematics in a contextualised and practical activity. In such a situation, cultural practices are considered known knowledge that should be used as a stepping stone to develop new mathematical concepts and theories. This applies ethnomathematical and humanistic perspectives to learning.

As defined previously in this chapter, numeracy is the ability to use language, numbers, images, and computers, as well as ability to communicate, gain useful knowledge, and apply dominant symbol systems of a culture (D'Ambrosio 1997). This broad definition makes it possible to use geometry topics to illustrate the processes of learning envisioned in the proposed pedagogy. Geometry provides examples of a visible mathematics that is physical in nature, and is portrayed in the Setswana culture as practical mathematics, more hands-on than theoretical. Learners acquire geometry by being involved in an activity. One such an activity is the construction of a traditional circular hut, pictured in Figure 12.2.

These types of structures were very common in the past, but still exist today, especially in villages. The building of this type of a shelter (circular mud hut) is used to demonstrate that mathematics and specifically here, geometry, has been successfully learned and used outside the formal schooling environments. The way to construct a traditional hut is something that almost all adults in Africa and particularly in Botswana are familiar with. In this example, the construction of the hut is used to teach the mathematical concept of a circle and its properties.

Every hut has a foundation. To make the foundation of a hut requires that builders mark the centre with a wooden peg. The builders then connect another peg by a string to the peg at the centre to construct a circle of preferred size, as indicated in Figure 12.3. What then emerges is a circular foundation that represents the circumference of a circle. The string between the marking peg and the stationary centre peg represents the radius of the circle. Conceptual knowledge of the circumference of a circle understood as a set of points

FIGURE 12.2 Example of a circular Setswana traditional hut (source: personal photograph of a traditional hut on University of Botswana Campus)

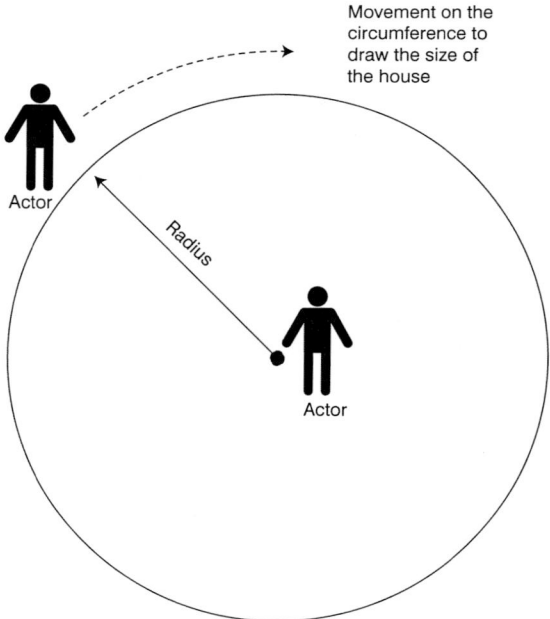

FIGURE 12.3 Making a circular foundation

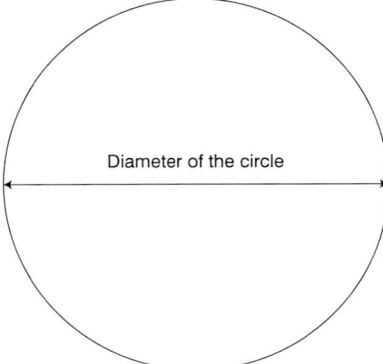

Diameter of the circle

FIGURE 12.4   Concept of a diameter

equidistance from the centre can be realised by showing that the peg at the end marked points which are equidistance from the centre because of the length of the string, which is constant. The concept is not immediately visible, but it can be defined at this point.

The diameter can be defined as the distance from one point on the circumference to the other opposite point through the centre peg as in Figure 12.4. Learners can formulate a relationship that the diameter is twice the radius since the distance from the point on the circumference to the centre (the radius) is the same for both points (one on the opposite side).

Figure 12.4 illustrates that the diameter is twice the radius. This illustration gives both the conceptual understanding (theory) and practical understanding of how the diameter can be represented. The length of the string can be varied to show that it is still true for any constructed circumference for the diameter to be twice the length of the radius.

The relationship between the circumference and the radius or diameter can be investigated by finding out how many times the string length makes up the circumference, which is likely to be approximately six times the radius or three times the diameter. This is yet another example of building a theory. Learners get to understand that the circumference is approximately three times the diameter or six times the radius. Having illustrated these dimensions, a life problem-solving scenario can be posed for learners to solve, for example, 'What length of the borderline of the hut is made from string length of three metres?' The solution to this problem can be extended to find out how much of the wall material is needed for the hut of a particular height. The wall brings in conceptual knowledge of a cylinder – a circular prism – a wall is built around a circular base foundation. From this observation, the wall and the two circular bases ends can be used to describe a cylinder. If the illustration is clear, it can even allow learners to make calculations regarding the surface area of the wall and the volume surrounded by the wall in a very concrete manner.

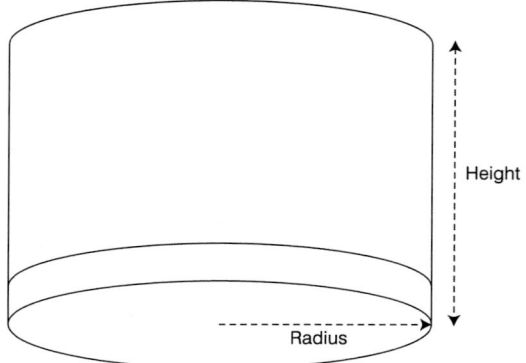

FIGURE 12.5   Calculations on the cylinder and relationship with a rectangle

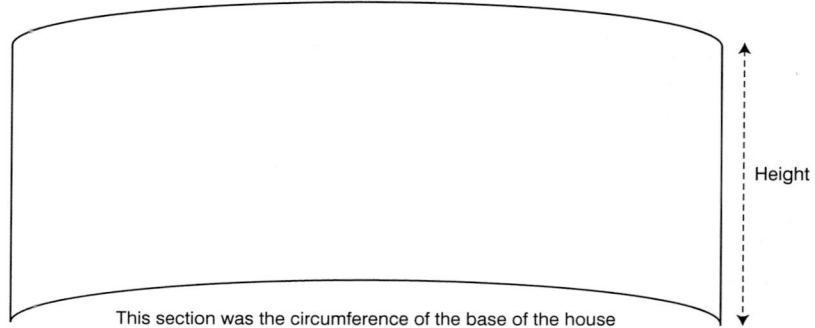

FIGURE 12.6   Base of a hut circumference

As illustrated in Figures 12.5 and 12.6, it can be demonstrated how the cylinder can be opened up to introduce the concept of surface area, and more importantly, the relationship with the side of the cylinder. Fully opened up, the cylinder will become a rectangle and the area can be calculated the same way as that of a rectangle.

The roof of the house, which is essentially a cone, can also be introduced to the learners to extend their mathematical literacy horizons. The concepts of calculation of area of a cone and its relationship with the amount of materials needed for roofing can then be worked into the mathematical skills. Since, at this juncture, people involved with this task would be expected to discover or derive the area of the base circle of the hut – area of the base circle of the hut ≈ 3 × radius × radius – the area of conic section can also be derived with its relation to the area of the circle. The conical shape can be effectively simulated to be formed from a part of a circle as Figure 12.7 demonstrates. Using this figure, if a larger part of the area of the circle is cut out along the line labeled '*s*' and rolled out by joining points A and B, it would look something like the conical shape

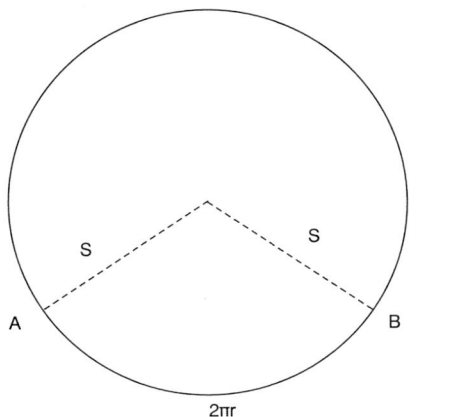 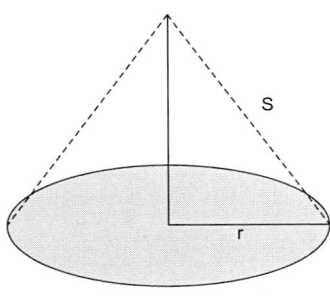

FIGURE 12.7   The roof concept

on the right. (The cone was cut along the lines (2*s*); the length of this cut is the slant height of the cone.)

   The area of the conical part of the hut is, in effect, the area of the larger part of the area of the circle on the left. With further demonstration, it can be shown that the area of the conical shape on the left is approximately 3*rs*, that is, three times the products of the radius of the circular cone base and the slant side of the cone.

   The case of the well-known traditional African practice of the construction of a hut has been presented to demonstrate that it is possible to approach the teaching of mathematics from the known to the unknown. The example given can be considered a case in which mathematical knowledge is invisible or hidden and needs to be unfrozen or discovered. In the process, concepts and theories are constructed and learners are prepared to use them to solve related real-life problems. This type of process uses Kane's (2002) suggestion of three types of numeracy, namely, 'constructible numeracy', 'visible numeracy', and 'useable numeracy'. Kane uses the term 'constructible numeracy' to indicate that learners are provided opportunities to build on and transform previous knowledge and ready themselves for problem solving. On the other hand, the idea of 'visible numeracy' illustrates the use of visual thinking and learning, while 'useable numeracy' relates to application. Kane's three concepts have been illustrated in the example of the construction of a Setswana traditional round hut.

   Another relevant term is 'invisible mathematics' as defined by Coben (2000). She describes this form as mathematics that a person can use but does not recognise as mathematics, similar to the way in which some people, described by others and by themselves as 'illiterate', engage in literacy practices which they do not recognise as 'literacy' (see Nabi et al., 2009). Most of the mathematical ideas embedded in cultural practices or events are in this form, hidden or frozen. When mathematical teaching and learning is approached from a cultural practice perspective, learners are confronted with discovering and unfreezing mathematical knowledge embedded in their respective cultural practices. In a situation like this, the task of the facilitator is to help learners unfreeze their hidden mathematical knowledge and to recognise this as their lived experiences. Facilitators who are successful in making the hidden mathematics knowledge more visible to learners help learners discover their deep wealth of mathematics knowledge, which, in turn, builds their numeracy confidence. When learners discover that they have rich experiences as a basis for learning, there is a great chance of success in making mathematics more meaningful and useful.

The case of the round hut example above addresses some major principles of adult learning and perhaps learning in general. One such principle is that learning is facilitated when lived experiences are taken into consideration. The mud hut, for example, presents a certain type of visible/physical presence of structure (round shape, reeds for roofs, wooden window, thatched top, mud wall) that learners in Africa are familiar with. This vividness allows learners to re-live their experiences of this structure whilst at the same time gathering new information, especially in relation to how such a structure embodies mathematical knowledge like geometry. This kind of visual perception allows learning to move from the known to the unknown. Learning unfolds in ways that are not abstract or random but concrete.

The case of the round hut presented in this section demonstrates that knowledge is socially constructed, although learners may not have been aware of this pattern prior to being presented with a familiar case to reflect on and learn from. Presentation of a case like that of the round hut also 'counteracts the generally accepted idea that all learners should learn the same academic subjects, which is part of the notion of a common school and common education for everyone' (Kincheloe 1999, 99).

Transformative numeracy pedagogical framework

Based on the principles demonstrated in the case provided above, a transformative numeracy pedagogical framework is presented for teaching adult basic mathematics like numeracy and geometry. Generally, transformative pedagogies are meant to actively engage learners using interactive techniques like cases, reciprocal dialogue, problem-solving scenarios and critical reflection. Greater focus on this type of pedagogy is mediated by two main principles. The first principle is that learning is meant to build functional skills. Mathematics knowledge, for example, is used every day in activities such as counting, weighing, measuring, ordering and sorting. These tend to be done in different ways by different cultures/societies.

The other principle is that of transformation. The target group for the proposed pedagogy are adults doing basic adult education. Characteristically, some of them have dropped out of school. Also, the length of being out of school differs. Some of these adults might have dropped out of school having had difficulty passing mathematics and thus have developed negative attitude towards this subject. Transformation for them would mean changing their mind-set, because, as Lawrence (2000) observed, it is common for negative attitudes that started when students first encountered failure in school to persist in adult life. For these adults, there is a need for re-skilling and up-skilling, especially to develop a positive attitude towards lifelong learning of numeracy so as to improve its use in everyday life activities.

Figure 12.8 illustrates the type of transformative learning envisioned in this chapter. One basic goal of transformative pedagogies is to ensure that learning results in improved self-image. Thus, the figure starts by acknowledging that each learner would come with different levels of mathematics understanding and use – the stage called 'Numeracy Now'. Some, as the figure indicates, may have no formal learning experience, whilst others may have some experience of learning mathematics in a formal school setting, but, perhaps, hold negative attitudes as a result of having not done well at the time. It is, therefore, necessary for the facilitator to find out about the backgrounds of the learners. One goal of transformative pedagogies is to ensure that learning results in improved self-image.

It has been observed that self-confidence normally leads to improved self-image (Lawrence 2000). Lack of confidence can result from a number of factors including pressure to cope with learning weaknesses especially in relation to learning a skill that society values, like numeracy (Schlöglmann 2006). It is the duty of facilitators to ensure that negative attitudes are replaced with positive ones so that learning can unfold without inhibitions. In the context of national literacy programmes, adult educators can inspire and motivate their learners largely through interactive learning approaches. These approaches can give practical knowledge of how numeracy skills can be applied in real-

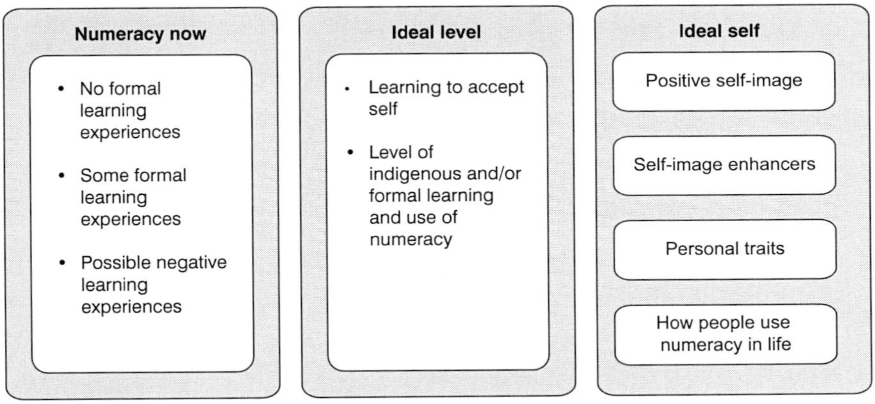

FIGURE 12.8   Transformative passage example

life situations. The ultimate goal of the transformative passage illustrated in this section is to ensure that numeracy skills are used for the betterment of people's life.

Recommended transformative pedagogy for teaching adult basic numeracy

Having outlined and explained the transformative passage envisioned in the learning of mathematics, a recommended transformative pedagogy for teaching adult basic numeracy is now presented.

Figure 12.9 gives the main characteristics of the pedagogy recommended for use with adults who aspire to learn mathematics, particularly numeracy/geometry at a basic level. It illustrates a road map/practical guide to teaching

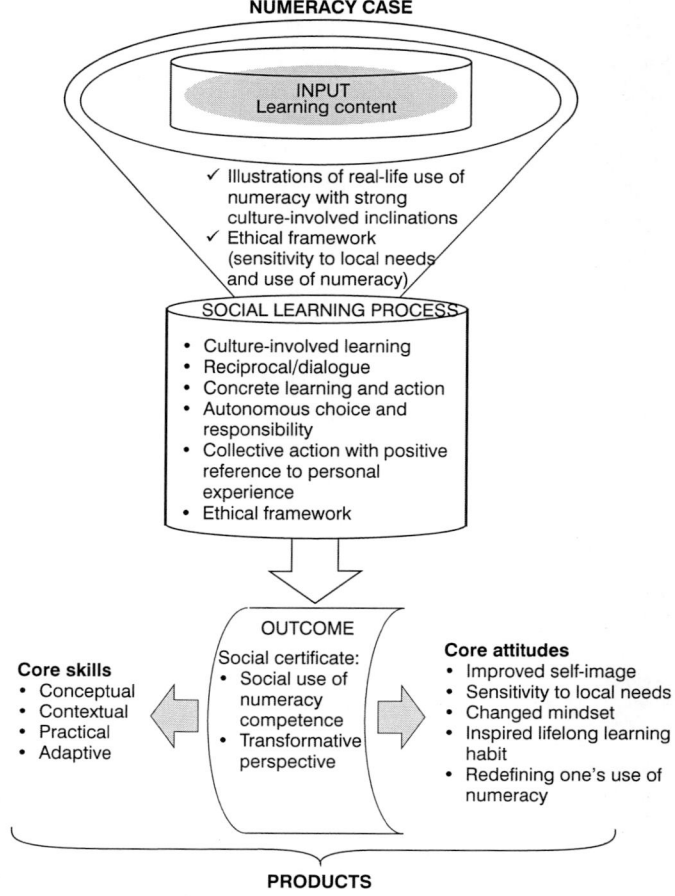

FIGURE 12.9 Transformative pedagogy adult basic numeracy

adult basic numeracy as a social-empowerment exercise. Social empowerment relates to the need to develop a socially numerate citizen. Ideally, such a person can use mathematics for personal benefit and also contribute to the lives of others. Developing a socially numerate person is an intricate process that can only be achieved when certain conditions are fulfilled as Figure 12.9 indicates.

Numeracy case

The caption, 'numeracy case' is placed at the top of the figure to indicate that a transformative pedagogy largely uses a numeracy case as a technique to help learners understand what is being taught or discussed. It is a case because it presents examples of real-life use of numeracy knowledge. By its nature, numeracy case has a strong culture-involved learning. Through illustration of building a Setswana circular traditional hut (Figure 12.2) a case of numeracy with strong culture-informed learning is presented. Culture-informed learning means that examples are packaged in a form that embeds mathematical knowledge, skills, and attitudes from a particular cultural practice. This type of case holds great meaning or significance for learners because of its familiarity. A case study also gives learners opportunity to link new knowledge to familiar situations. In the context of the NLP, the facilitator becomes a guide who prepares the case to be discussed and ensures that learning activities are appreciated for their practical value. In short, there are some basic conditions to be followed to drive a successful transformative pedagogy for adult basic numeracy.

Social learning processes

It has been indicated that transformative pedagogy for adult basic numeracy is a social learning process. To qualify to be a social learning process means that this pedagogy makes use of and promote social skills such as reciprocal dialogue, autonomous choice, ethical values and decision-making. These aspects can drive effective learning of basic numeracy for adults as explained below.

Reciprocal dialogue

Transformative pedagogy promotes the idea that learning is accomplished mostly through communication and reciprocity. Communication is viewed informally as an expression of ideas and actions which take on a form of meaningful interactions in a particular learning set-up. There are different contexts of adult learning, each with its unique setting, thus what is considered meaningful communication in one setting may differ in another, and may be dependent on such characteristics as composition of learners, resources available, and location of learning. The specific type of communication will arise from how these characteristics work together to promote relevant learning. Reciprocity or dialogue, for example, is an integral aspect of effective learning. Reciprocity recognises that human beings can learn from each other. This is consistent with other theories of learning such as constructivism and theories that recognise learners as social beings.

Reciprocity is also acknowledged as promoting real-time interaction between facilitators and learners and/or among learners themselves. Active interaction is expected between human and non-human aspects in the learning environment such as learner and learning resources. Proper learner engagement and interaction is promoted when all aspects are synchronous or when each aspect mutually influences the other (Wagner 1994). As Jensen (2008) points out, meaningful learning takes place when learners mutually adapt their behaviours and actions to the learning environment. Reciprocity in the learning environment is thus one of the guideposts to know that relevant learning is taking place.

Autonomous choice

Much of the literature on adult learning assumes that adults are goal-oriented. Their choice to engage in learning is a result of what they believe such learning would contribute to their lives. Adult learning is, then, for the most part purposeful and adult learners' motivation would be high when they are given freedom to indicate their preferred learning contents and methods. Though this principle is not religiously adhered to in all adult learning situations, its widespread use can promote a learner-centred approach and self-directed learning. It may also help to resist the overwhelming use of contents divorced from learners' needs. There are people who believe that given the chance, adult learners can have full control over their own learning (Birkenholz 1999).

Ethical values

Learning does not bracket the cultural values learners bring to the learning environments. Rather, these values determine how learning is engineered. It is, therefore, vital that both learners and facilitators should understand the ethical framework that may influence what is learned and how. Ethics may be acted out in learning processes, for example, through interactions and relationships in the learning environment. But as Chia (1998, 377) says, the behaviour displayed during learning 'is a larger conceptual and social category than the execution of a task'. Chia is right to say that there will always be social values and significance in the way people behave and act. Thus viewed, an ethical framework can influence both the content and processes of learning.

Decision-making

Learners are provided with opportunities for decision-making, especially in deciding what to learn. When learners are allowed to make autonomous choices, they learn what will benefit them in real life, and this is never divorced from their lived experiences. Transformative pedagogy does not push out the lived experiences of the learners; hence, reference is made to culturally responsive learning. Concrete learning is also promoted when learners are allowed to reflect on their lived experiences and use these reflections to change negative assumptions and promote positive experiences of learning numeracy.

Outcomes: core skills and attitudes

The third major aspect of the recommended transformative pedagogy is 'Outcomes'. Traditionally, programmes of learning lead to the awarding of certificates. In the formal school setting, for example, certificates of accomplishments are awarded to those that have satisfied the requirements of course in terms of grades achieved. Critics of these certificates, like O'Sullivan (2006), contend that, although these certificates are said to be passports to good jobs and positions of responsibility, such as leaders and managers, this is not always the case. The phenomenon of educated unemployment, for example, bears evidence that being educated does not assure a person a job. In Botswana, educated unemployment among youth keeps rising. Epstein and Yuthas (2012) consider academic certificates Western ideas that are ill-suited to the needs of some people in the African context, especially the rural populace in places like Botswana. Searle (1981) warns that education becomes a true asset if what happens in institutions is never cosmetic, but can be applied to the social and political conditions present outside the classroom walls. The term 'social certificate' might be more suitable to be used here because it underscores the importance of learning experiences that imbue learners with skills to use in their respective contexts; cities, towns, urban villages, rural villages and settlements as is the case in Botswana. The concept of a social certificate is further elaborated in the discussion which follows.

Social certificate

Social certificate is a concept which denotes the need to develop a 'socially numerate person'. As defined previously in this chapter, a socially numerate person is someone with practical knowledge and self-confidence to act individually and collectively in using numeracy skills to address real-life challenges (Ernest, 2010). The term social certificate thus denotes achievement of values or skills of how to live in real worlds. By dialoguing and freely sharing their experiences, learners emerge with practice-oriented behaviours and values widely accepted for survival and improved livelihood. The concept of social certificate thus implies that people can act on their life situations using what they learned.

Figure 12.9 outlines four main skills of conceptual, contextual, practical and adaptive skills including attitudinal areas such as improved self-image, sensitivity to local needs and use of numeracy, a changed mind-set about the usefulness of numeracy, and an inspired lifelong learning or learning-to-keep-learning attitude. These are the core functional skills of a transformative adult numeracy pedagogy suggested in this chapter.

In this context, conceptual understanding relates to knowing basic numeracy concepts and other textual numeracy information and explaining these concepts as they relate to their application to real life situations. This kind of understanding is essential in applying numeracy to real life activities. Practical is synonymous with numeracy proficiency, indicating the extent to which the skills of basic numeracy gained are applied to real life situations. Application is also a direct measure of contextual understanding. It relates to the ability of a person to interpret, analyse and apply numeracy skills in different social contexts. Adaptation acknowledges that adults have different roles in the communities and these may call for application of numeracy skills, thus an adaptive person will know how to use the skills she/he has to suit each situation. For effective application, these core functions of numeracy skills should be merged with proper attitudes such as improved self-image, a positive mind-set about the usefulness of numeracy, and an inspired lifelong learning attitude. This type of basic level of numeracy is essential for full participation of adults in their roles in communities (OECD 2012).

Summary

In this chapter, a case-study approach to learning mathematics is recommended. A case approach is, to a great extent, a way to fill in the void left by past negative experiences of prescriptive learning formats of formal mathematics. Using Kincheloe's (1999, 99) words, one can describe the model presented in this chapter as a pedagogy that aims to 'train the mind by training the hand' because conceptual and practical knowledge are intertwined. The model is intended to present content in ways that limit the prescriptive and memorisation nature of formal classroom learning of numeracy. The separation of conceptual and practical understanding is not logical, as adults are not in class to get grades, advance to another class, or get employment, but to acquire functional skills.

Furthermore, adult learning privileges discourse where 'students interact with subject content, transforming and discussing it with others, in order to internalise meaning and make connections with what is already known' (Nicol and Macfarlane-Dick 2006, 200). The interactions transform the way learners think about the value of learning numeracy. A numeracy case is recommended because it provides ample opportunities for interactions. New forms of thinking and new patterns of using numeracy are expected to be developed through different types of interactions experienced. Using a numeracy case is in fact a test to see how what is learned is still embedded in societal practices, which in turn makes learners appreciate their experiences as members of certain communities. As interacting agents in the learning environments, learners get to value interactions as sources of their livelihood, just as much as they cannot hope to survive without continuous learning.

References

Aloni, A. (1999) "Humanistic education," *Encyclopaedia of Educational Philosophy and Theory*   http://www.mofet.macam.ac.il/prof/dialog/Documents/aloni%20ency%20 humanism.pdf (accessed 14 December 2016)

Ascher, M. (1991) *Ethno-mathematics: A Multicultural View of Mathematical Ideas* London: Cole Brooks

Babungura, A. K. (2002) "Infusing ethno-mathematics/ethno-science in the curriculum," paper presented at the Symposium of African Universities in the 21st Century, University of Illinois, Chicago, 20 July

Birkenholz, R. J. (1999) *Adult Learning* Crete, IL: Interstate Publishers

Chia, R. (1998) "From complexity science to complex thinking organization as simple location," *Journal of Organization: The Interdisciplinary Journal of Organization, Theory and Society*, 5(3): 291–393

Coben, D. (2000) "Numeracy, mathematics and adult learning". In Gal, I. (ed), *Adult Numeracy Development: Theory, Research, Practice* New York: Hampton Press

Davis, J. (1993) *Better Teaching More Learning: Strategies for Success in Postsecondary Settings* American Council on Education, Series on Higher Education, Phoenix, AZ: Oryx Press

D'Ambrosio, U. (1985) "Ethnomathematics and its place in the history and pedagogy of mathematics" *For the Learning of Mathematics*, 5(1): 44–48

D'Ambrosio, U. (1997) "Ethnomathematics and its place in the history and pedagogy of mathematics". In Powell, A. and Frankenstein, M. (eds), *Ethnomathematics, Challenging Eurocentrism in Mathematics Education* Albany, NY: State University of New York Press

Ellerton, N.F. and Clements, M.A. (1990) "Culture, curriculum and distance teaching of mathematics". In Evans, T. (ed), *Research in Distance Education*, Geelong VIC: Deakin University Press

Epstein, M. and Yuthas, K. (2012) "Redefining education in the developing world" *Stanford Social Innovation Review*, 10(1): 19–20

Ernest, P. (2010) "The social outcomes of learning mathematics: Standard, unintended or visionary?" http://research.acer.edu.au/cgi/viewcontent.cgi?article=1090&context =research_conference (accessed 23 April 2017)

Glevey, K. (2009) "Pupils of African heritage: Mathematics education and social justice". In Enerst, P., Greer, B. and Sriraman, S. (eds), *Critical Issues in Mathematics Education* New York: Information Age Publishing

Jensen, E. (2008) *Brain-based Learning: The New Paradigm of Teaching* Thousand Oaks, CA: Corwin Press

Kane, M. (2002) "Validating higher stake testing programs in educational measurement" *Issues and Practice*, 21(1): 31–41

Kincheloe, J. L. (1999) *How Do We Tell the Workers? The Socioeconomic Foundations of Work and Vocational Education* Boulder, CO: Westview Press

Knowles, M.S. (1984) *Andragogy in Action: Applying Modern Principles of Adult Learning* San Francisco, CA: Jossey-Bass

LawNerds.com (n.d.) "Gets smart about the case method" *www.lawnerds.com/guide/ briefing.html* (accessed 13 December 2014)

Lawrence, D. (2000) *Building Self-esteem with Adult Learners* London: Paul Chapman Publishing Ltd, London

Lekoko R. N. (2015) "Preface". In Lekoko, R. (ed), *Cases on Grassroots Campaign for Community Empowerment and Social Change* Hershey, PA: Information Science Reference

Malloy, C. (2008) "Looking through the world for democratic access to mathematics". In English, L. (ed), *Handbook of International Research in Mathematics Education* New York: Routledge

Mbiti, J. S. (1988) *African Religions and Philosophy* London: Heinemann

Maruatona, A. and Mokgosi, A. (2006) "A national literacy programme: A situation analysis of evaluation practices" paper presented to the Cross-National Workshop on Reinforcing National Capacities to Evaluate NFE and Literacy Programmes for Young People and Adults, hosted by UNESCO Institute for Education, Hamburg, Germany 20–23 February

Millroy, W. L. (1992) *An Ethnographic Study of the Mathematical Ideas of a Group of Carpenters Journal for Research in Mathematics Education Monograph No 5* Reston VA: National Council of Teachers of Mathematics

Mpofu, S. and Youngman, F. (2001) "The dominant tradition in adult literacy: A comparative study of national literacy programmes in Botswana and Zimbabwe" *International Review of Education*, 47(6): 573–595

Nabi, R., Street, B. V. and Rogers, A. (2009) *Hidden Literacies: Ethnographic Studies of Literacy Practices in Pakistan* Bury St Edmunds: Uppingham Press

Nicol, D. and Macfarlane-Dick, D. (2006) "Formative assessment and self-regulated learning: a model and seven principles of good feedback practice" *Studies in Higher Education*, 31(2): 199–218

Nyerere, J. K. (1973) *Freedom and Development* Oxford: Oxford University Press

OECD (2012) *Literacy, Numeracy and Problem Solving in Technology-Rich Environments: Frameworks for OECD Survey of Adult Skills*, Paris: OECD Publishing

O'Sullivan, M.C. (2006) "Lesson observation and quality in primary education as contextual teaching and learning processes" *International Journal of Educational Development* 26: 246–60

Podsen I. (2013) *Teacher Retention: What is Your Weakest Link?* New York: Routledge

Powell, A. B. and Frankenstein, M. (eds) (1997) *Ethno-mathematics: Challenging Euro-centrism in Mathematics Education* Albany, NY: State University of New York Press

Rosa M. and Orey, D. C. (2011) "Ethnomathematics: The cultural aspects of mathematics" *Revista Latinoamericana de Etnomatemática*, 4(2): 32–54

Schlöglmann, W. (2006) "A lifelong mathematics learning: a threat or an opportunity? Some remarks on affective conditions in mathematics courses" *Adult Learning Mathematics: An International Journal*, 2(1): 6–17

Searle, C. (1981) *We Are Building the New School! Diary of a Teacher in Mozambique* London: Zed Press

Sheffield, J. (1972) "Non-formal education in Africa: Micro-solutions to macro-problems?" *African Studies Review*, 15(2): 241–254

Statistics Botswana (2014) *Statistics Botswana Annual Report 2013/14*. Gaborone: Government Printers.

UNESCO (2011) "Adult basic education programmes: Country Profile, Botswana" http://litbase.uil.unesco.org/?menu=9&programme=96 (accessed 12 December 2016)

Wagner, E. (1994) "In support of a functional definition of interaction" *The American Journal of Distance Education*, 8(2): 6–29

13

THE WORKPLACE AS A SITE FOR LEARNING CRITICAL NUMERACY PRACTICE

Keiko Yasukawa

Workplaces as research sites for numeracy as social practice

The workplace as an important site of numeracy practices has been the subject of much research (see for example, Fitzsimons 2005; Hoyles, Noss, Kent and Bakker 2010; Kane 2014; Kanes 2002; Wedege 2000; Williams and Wake 2007; Yasukawa, Brown and Black 2013; Zevenbergen 2011). Much of the practice-focused research of workplace numeracy and mathematics is based on *in-situ* studies, involving close observations of mathematical activities in the workplace supplemented by in-depth interviews with the workers. They share as their starting point key insights from earlier studies (for example Baker 1998; Harris 1991; Lave 1988; Nunes, Schliemann and Carraher 1993) which established: first, mathematics in workplaces can look very different from the mathematics that is practised and valued in schools; second, mathematical practices in workplaces are not always recognised by the workers themselves as mathematical activities; and third, the 'transfer' of mathematical knowledge and skills learnt in schools to the workplace is not straightforward or unproblematic (see also Evans 1999). Thus, these workplace numeracy studies suggest there are interesting questions about the perceived and actual saliences of workers' previous mathematics education as they experience their working lives; they point to the need to take a nuanced examination of the apparently indubitable importance of mathematics education in people's lives.

Within the workplace, researchers have uncovered the politics of knowledge, that is the unequal legitimacy and visibility attached to the mathematics in the school curricula relative to mathematics in workplaces, particularly when the workers doing the work are characterised as low-skilled. In addition, research has uncovered different kinds of power relations surrounding workers' numeracy practices. These include the gatekeeping by workplace supervisors of information about the mathematical algorithms that workers are tasked to implement (Williams and Wake 2007), the inaccuracies of the popular deficit discourses about young workers' numeracy practices (Zevenbergen 2011), the enduring affective impact of school mathematics experiences on adult workers' self-perceptions as do-ers and users of mathematics (Wedege 2002), and 'myths' surrounding crisis discourses about low levels of workers' literacy and numeracy skills (Yasukawa et al. 2013). By challenging orthodoxies surrounding mathematics or numeracy as 'basic' or 'foundational' skills that can be contemplated in isolation of their context of use, these practice-focused studies of numeracy have contributed critical insights about the political dimensions of workplace numeracy practices.

This chapter focuses on numeracy learning among groups of workers, examines the possibilities and limitations of workplaces as sites of learning *critical numeracy practices* and considers their implications for education for work. The human capital-based discourse dominating many policy discussions about lifelong learning places significant emphasis on literacy and numeracy as key skills for economic growth and productivity, and for individuals' successes in the globalised labour market (see for example OECD 2013). While formulations of numeracy in many influential policy discussions, such as the OECD's Programme for the International Assessment of Adult Competencies (PIAAC) initiative, acknowledge numeracy as necessarily contingent on the context in which it is used, the discussions do not generally engage with issues of power that impact on the adults as employees and what they are able to negotiate about their place in the workplace. The chapter explores the possibility of workers developing critical numeracy practices: understanding, questioning and perhaps even mounting a challenge to the assumed positioning of workers in their workplace. It does so with the aid of the theoretical resources of Engeström's (2001) third generation cultural-historical activity theory (CHAT).

Politics of education for the workplace

Yasukawa and Brown (2012) posited a taxonomy of 'mathematics for, and in, the workplace' that had four categories of the purposes to which mathematics is put to use in the workplace: 'enabling mathematics for accessing training and qualifications'; 'technical mathematics for doing the job'; 'functional mathematics for being a paid worker' (for example, checking one's pay, completing tax returns, filling in timesheets); and 'critical mathematics for "reading" the politics of work and the workplace' (Yasukawa and Brown 2012, 252). They suggested that their fourth 'critical mathematics' category is often neglected in discussions about workplace mathematics and numeracy, but include mathematics-imbued activities such as:

- Analysing the 'logic' of the pay rates in the workplace
- Questioning data on workplace 'productivity' and/or link to pay

- Analysing the political representation of different groups of workers [in their workplace]
- Analysing the political representations of different groups of workers in the union.

(Yasukawa and Brown 2012, 252)

To understand the absence of this critical dimension in many workplace education and training programs, insights about the different objects of learning in and for the workplace are helpful. Working with Engeström's (2001) third-generation cultural-historical activity theory (CHAT) as a lens, the labour educator Helena Worthen identifies two distinct knowledge-producing activity systems in workplaces:

> Each produces knowledge about how to do the work of that workplace, but they are differently motivated: one toward productivity, and the other toward earning a living.

(Worthen 2008, 322)

Focusing on *activity systems* as units of analysis is a characteristic of Engeström's (2001) CHAT. Engeström built on Vygotsky's original theorisation of learning and development as a goal-oriented activity, extending Vygosky's individual learner focus to a team-focus given that in many workplace and other organisational contexts, learning is a team-based activity. Engeström's formulation provides a way of viewing the learning from the perspective of the learning *subject(s)*. The subject learns within an *activity system*, which consists of: the subject's *object* or motive for learning; the *mediating tools* or instruments that support the learning; the *rules* including the custom and practice of the *community* in which the subject belongs; and the system of *division of labour* that is in place that impacts the subject's involvement in the activity (Engeström 2001).

Worthen coins the terms 'work process knowledge' and 'negotiating knowledge' to distinguish between the two kinds of knowledge pertaining to the activity systems whose objects are productivity and earning a living, respectively. In policy discourses, it is often the former rather than the latter motivation that is used to promote adults' numeracy learning. For example, Yasukawa and Black (2016) showed by tracing the development of Australia's National Foundation Skills Strategy for Adults, that the impetus for the policy came from fears felt by industry and employer groups that workers' numeracy and literacy 'deficits' were holding back productivity. Not surprisingly, these stakeholders did not raise concerns about poor numeracy being a barrier to workers' ability to negotiate better working conditions or wages.

The unfolding of the politics of workplace literacy and numeracy education has been illuminated in many early critical studies of literacy (see for example Belfiore, Defoe, Folinsbee, Hunter and Jackson 2004; Hull 1997). They point to the demand of changing work process knowledge as the driver for workplace education and training, and not surprisingly so, given that most of these programs are initiated by the employer. Many of these programs are often justified by a discourse of literacy and numeracy deficits, especially among low-paid workers (Hull 1997). Moreover, most workplace literacy and numeracy programs, even if they are not funded by the employers themselves, are secured through the initiative of, or at least the agreement of the employers. Gallo (2004) sums up the conflict this presents for the educator:

> workplace educators find themselves caught in the midst of political and moral dilemmas, as they struggle to frame their work in a way that appeases those who are paying for classes and appeals to those that are participating in them. While many employers claim to be seeking a more highly skilled workforce ..., they are often reluctant to offer the greater wages commensurate with such skills. Instead, by providing on-site literacy instructions to their employees, companies proposed to improve the skills of their current workforce as well as productivity and profitability.
>
> *(Gallo 2004, 1)*

Gallo (2004) gives an account of her own worker-centred literacy programme that did, at least to some extent, enable the women English-as-a-second-language (ESL) workers in the workplace to develop what Worthen would call negotiating knowledge. Gallo explains that by allowing learners to have a voice about how they are defined as learners, for example by their potential and aspirations rather than by what they lack, she was able to 'create pockets of community and hope' that helped workers to examine the power structures and communicative practices in their workplaces with a view to improving them, rather than simply accepting them (Worthen 2004, 131). She provides examples of how the workers learned to make a case to their employer about improvements they were seeking by recasting these improvements as changes that would carry benefits also for the organisation.

The importance of creating 'pockets of community of hope' is a factor that Bond (1999) also identified when she was faced with a group of men whose negotiating knowledge pertaining to their livelihoods had been put to the ultimate test in a prolonged period of industrial action. Bond (1999) recounts how, as an outsider to a group of striking sheet-metal workers on a picket line, she had first to be accepted into the workers' community before she could fully recruit them into her research on men's non-participation in return-to-learn courses. She gained their trust through demonstrating shared political orientations and her links with people who were known and respected by these workers. As a researcher and educator whom the men regarded as 'one of them', Bond was able to learn about their perspectives on learning including what would motivate them to participate in learning and what they found as obstacles. Thus, both Gallo (2004) and Bond (2000) recognise that the affordances for learning are not rendered visible to the workers in the absence of a pedagogical relationship between the educator and the workers that are affirming of the knowledge that the workers bring to the learning, and which are built towards outcomes that have hope for improving their understandings about, and impacting upon, their existing situation as workers.

In his critique of the popular deficit model of workers' education that attributes blame on adults for struggles they have in securing employment, Jacobson (2016) puts forward a case, not for greater training in literacy and numeracy or other skills that are claimed to improve employability, but for an education about the political economy of adult education, that is for teachers and their learners alike to consider the learners' individual experiences in relation to larger socioeconomic forces. He argues 'adult basic education should reject the rhetoric of workforce development and the illusory economic premises that it is based upon' (Jacobson 2016, 18). In short, Jacobson, along with Bond, are calling for a shift to a more radical stance among those who want to see adult education being a struggle for greater social justice.

In terms of Engeström's (2001) CHAT framework, a numeracy practice might be interpreted as an activity system that, from the worker's perspective, has a motive or object that leads to a positive outcome for the workers, and supporting resources or mediating tools that help the workers realise their goal (see also Kanes 2002). Gallo's (2004) and Bond's (1999) insights suggest that there would be a sense of worker solidarity and agency in this learning endeavour. Building on Vygotsky's concept of the zone of proximal development (ZPD), the metaphorical region within which an individual's knowledge could be extended through interaction with a peer or teacher, Engeström focuses on the kind of collective learning that results in a radical transformation of existing arrangements of an activity system as 'a collective journey through the zone of proximal development of the activity' (Engeström 2001, 137). He suggests that such learning is often provoked when the existing activity system is disturbed, for example, through the introduction of new technology, a new set of rules or policy about how work is to be conducted, or a restructure of the work group itself. This kind of learning, when it occurs, is called expansive learning (Engeström 2001). The kind of radical education advocated by Bond and Jacobson, and the kind of learning that occurs in what Worthen calls a negotiating-knowledge-producing activity system would likely fall into the category of expansive learning. If numeracy is invoked in a substantial way in such an activity system, the outcome of the activity system may be critical numeracy practices that contribute to a shift in power relations within the workplace in some way.

The concept of a collectively traversed ZPD that is focused specifically on numeracy development may be captured by the notion of a *numerate environment* posited by Evans, Yasukawa, Mallows and Creese (2017). Extending the work undertaken on the notion of a 'literate environment' by the European Union High Level Group on Literacy (2012), Evans *et al.* explored the idea of a numerate environment. They suggest that in order to understand how adults negotiate everyday numeracy activities, an ecological perspective that takes into account the adults in relation to their context of the learning would be useful. To this end, they propose a characterisation of what they call the *numerate environment* as one with opportunities, support resources (rather than barriers) and also positive demands for people to use and develop their numeracy skills. In contrast to this would be contexts that effectively deny people the opportunities for numeracy development by focusing on their deficits and/or not providing demands on the people to engage in numeracy practices. Like the ZPD, the numerate environment is dependent upon the presence of a meaningful goal or motive for learning and resources which could be people, symbolic tools or material resources to mediate the learning. Their concept emerged through their questioning of the deficit focus taken by commentaries on the numeracy results of the PIAAC, and the possibilities that the affordances that adults have for numeracy development need to be examined, rather than simply attributing blame on adults who do not achieve high results in surveys such as the PIAAC.

In the following section, two case studies of workplace numeracy learning and practices from the author's previous research (Brown, Goodman and Yasukawa 2006; Black, Yasukawa and Brown 2013) are revisited and analysed using CHAT resources to examine and highlight workplace contexts that are 'critically numerate' environments and those that are not. In one, the author was one of three activist researchers who had been researching the lived experiences of casual (hourly paid) academics in a university (Brown, Goodman and Yasukawa 2006, 2010; Yasukawa and Brown 2012), and in the other, one of three researchers investigating production workers' literacy and numeracy practices (Black, Yasukawa and Brown 2013; Yasukawa, Brown and Black 2013, 2014).

Learning and bypassing negotiating knowledge for critical numeracy: two case studies

Casual academics in the academy

In the Australian higher-education sector, employment of a large number of teaching staff on hourly-paid contracts has become accepted practice as growth in student numbers continually surpasses increases in the government's funding for the sector. This kind of hourly-paid employment is known in Australia as casual employment, and these teaching staff as casual academics. There is, however, nothing very 'casual' about their work. These academics are issued a contract that specifies the dates and hours of classes that they are contracted to teach for the whole teaching semester (normally between 12 and 13 weeks), and their pay for that semester is based on the number of hours of lecturing or tutoring they perform multiplied by the designated hourly rate for their teaching activity.

Casual employment of academics, for the universities, is a cost-effective way to staff the increasing number of classes, and the casual academics are hired on the basis of their academic and professional expertise in the area that they are teaching. However, the casual employment is insecure; no matter how many semesters a casual academic has taught the same subject, there is never any guarantee that they will be employed again in the following semester. There is no career path, no severance pay even if after 10 years of casual work the university decides not to re-employ them. With no job security, casual academics find it impossible to plan their future: they struggle to secure a mortgage; and they cannot plan and in some cases, afford any holidays. They have limited workplace rights and voice in the decision-making processes within their departments, even in relation to the curriculum and assessment that in many cases they know more about than the fulltime permanent academic who is the nominal coordinator. These and other stories of the lived-experiences of casual academics in one university surfaced in the research by Brown et al. (2006; 2010).

At the time of their research (Brown et al. 2006), there was very little opportunity, let alone support, for casual academics to develop their negotiating knowledge in the academy. The trade union for the Australian higher education industry, the National Tertiary Education Union (NTEU) had, until then, not invested a great deal of effort in recruiting and organising casual members. The NTEU leadership had been, and continues to be, dominated by university staff who hold secure positions, and the issues affecting casual academics have tended to take second place to other issues. This has also been exacerbated by the practical challenges of organising staff who are largely excluded from collegial structures of their place of work, and who have no offices of their own where they could be contacted. However, with the growth of casualisation of academic work in the sector, and the heightened awareness among all union members about the industrial injustices being experienced by their casual academic colleagues, the union started to take the concerns of casual academics more seriously. The stories of the casual academics' lived experiences became one of the tools for organising in some of the university branches (Yasukawa and Brown 2012).

After organising activities increased in one university, casual academics started to engage their union delegates to help address their concerns. Yasukawa and Brown (2012) provide an account of a casual academics organising campaign that led to the development of critical numeracy learning among a group of casual academics who contacted the union about their concerns about underpayment. Meetings of the concerned casual academics were organised by the union. All of these casual academics were engaged as tutors in the same department, and all were concerned that they were not being paid for the work they performed. As highly qualified academics, checking that the number of hours of tutorials that they taught and the tutoring rate were multiplied correctly on their contract was not a difficulty; however, the casual academics were mystified by how their employer appeared to be getting away without paying for the many hours of what they saw as unpaid class preparation, student consultations outside class and marking. What was not clear to them was how the 'fine print' surrounding the pay schedule in the university's enterprise agreement was to be interpreted:

> The hourly rate of pay for … tutoring … contained in Schedule 2 will encompass the following activities in addition to the delivery of … tutorials:
> a   Preparing … tutorials
> b   Up to 20 minutes of marking for each hour of teaching
> c   Administration of relevant records of the students for whom the casual employee is responsible and
> d   Consultations with students
>
> *(Yasukawa and Brown 2012, 259)*

As the union delegate, the author helped to demystify the intent of this model of payment; that the hourly rate was not in fact an hourly rate but a multiple hour-rate. As stated in another part of the enterprise agreement, the pay for one hour of tutoring is to be interpreted as pay for: 'one hour of delivery and two hours of associated non-contact duties' (Yasukawa and Brown 2012, 259).

Thus the two hours of 'non-contact duties' are assumed to encompass all of the preparation, student consultation and marking duties. What appeared on the surface as a reasonable good hourly rate suddenly became very ordinary when it was divided by three.

Once the casual academics understood for how many hours of work they were being paid in their contract, they realised the number of hours they were being expected to spend to do their work. A collective decision was then made for the casual academics to keep a log of the time they spent on various preparatory, marking and student consultation work, for the remainder of the semester. At the end of the semester, the tutors' data showed some, but not large, variations, and when the average of their data was compared to the number of hours their contract specified, it became clear that all of the tutors had been working many more hours than they were being paid for. An industrial dispute was lodged by the union, anonymised data from the members were tabled, and back pay was won.

Production workers in flexible manufacturing

Hearing Solutions

The second study focuses on a different group of workers: production workers in manufacturing firms. For over two decades, manufacturing companies have been applying what is known alternatively as lean production, flexible manufacturing or competitive manufacturing principles to improve productivity and remain competitive in the globalised economy (Black et al. 2013). In simplest terms, lean modes of production are aimed at producing 'more for less' under strict systems of quality control. During 2012, two companies, a customised hearing aid manufacturing company Hearing Solutions and an industrial gasket-producing company Insulation Products, were in the process of delivering lean production training modules to their production workers.

Hearing Solutions had already delivered the initial course to the workers and when the research took place, the workers were already applying many of the emblematic features of lean production: working in well-defined teams, each graphically tracking their productivity against daily, weekly and monthly performance targets on their team's business boards, and holding daily team meetings in the morning to review their performance against their productivity targets. The 'lean numeracy practices' of counting incoming work orders, outputs, hours worked, worker absences, and comparing these numbers against targets on bar graphs had become routine. At the time that the research team visited the firm, a group of team leaders was being trained for the next level qualification which would involve development of further 'lean numeracy practices'. In the training session, eight team leaders were being taught the concept and procedures of 'value stream mapping': costing each step of the production process, for example, by analysing the distance and time taken to transfer the different components of the product from one stage to the next in the assembly process.

During the training session, the trainer commented on progress made in one team, saying:

> You had some really good stuff on the timing, how long it took. It was eight minutes and now it's five or something, I can't remember exactly. So you guys have got some good, strong stuff.

Throughout the training session, the trainer continued to elicit workers' suggestions of improvement projects that the team leader had identified for their value stream mapping:

Team leader: I've got one, yes. Orders received each hour, to monitor that from the start.
Trainer: So you've before and after?
Team leader: Before and after.
Trainer: Good.

Another team leader identified mapping the cost from 'delivery to dispatch' and targeting improvements in the timing. One team leader talked about the efficiency his team made in the way they disposed of waste from the production process:

Team leader: we come up with that idea – I mean previously before we [came] up with that idea it would take us hours and it's so dirty and so

Trainer:

stink to clean that barrel of the waste. So since we think of that idea then now we have to throw the whole lot out and put a new thing in and we don't need to clean it and it saves us a lot of time. This is the sort of stuff we're looking for.

The training was focused entirely on identifying and reporting on the improvement projects that were underway, and then preparing the evidence they would need to present to the assessor so they could obtain their qualification. In the separate individual interviews with the team leaders and other production workers, there was little resistance to the training or to how their work practices were being transformed; most saw benefits for their work process, and some were highly enthusiastic about how this mode of production could improve quality and customer satisfaction. The work process knowledge that was generated from the training was evident from both what the workers said, and what could be observed on the production floor. However, there was no comment in the training session or in the interviews that suggested connections were being made between the improved productivity they were achieving for the company and what it could mean for improving the workers' earnings.

When asked about whether the improvements led to any changes in pay or working conditions, all except two team leaders had little to say. While not being hopeful of any change, one said that a 'downside' of the company was that the workers were not seeing improvements in 'the financial stuff'. Another team leader indicated greater insight into the possible barriers to this, citing the national industry award that sets the minimum pay and conditions for workers in the manufacturing sector, and noting that because most of the workers had been employed at Hearing Solutions for a long time, they had all reached the top of their pay scale. What was not being questioned by the workers in the interviews was whether there were avenues for negotiating for better pay and conditions than the minimum award and the unrealised negotiating power that they held as producers of the company's goods.

Insulation Products

At Insulation Products, the workers were just being inducted into the principles of lean production when the research took place (Black et al. 2013). The trainer elicited the different problems that workers experienced in performing their work efficiently. She started by offering examples such as the time taken to locate or access materials they needed, waiting for the delivery of materials and 'double handling' of work processes. Several workers were forthcoming with examples of problems in their workplaces, including awkwardly placed equipment that made it time-consuming to access it, and excess material or scrap materials not being processed, thus creating what a worker referred to as 'the mess' on the production floor.

The trainer then moved onto what improvements could be put in place to address the problems that the workers had identified. It is in the exchanges that followed where the stark difference between activity systems that generate work process knowledge for the company and negotiating knowledge to improve the life of the workers can be seen. Two of the workers explained that 'the mess' was a consequence of the company not replacing the worker whose primary role was to sort off-cuts and to store unused materials where they belonged; they also said that the storage space was tight. They said that filling this position would solve much of 'the mess' problem. However, the trainer was quick to dismiss this as an acceptable solution: replacing the lost staff would not be economically viable, and she had assessed the storage space to be quite adequate. The trainer then asked the workers to identify changes they could make to address the problem. One of the workers commented that there was insufficient care among the workers. This then led to the following exchange between the trainer and the worker:

> Trainer: What inspires care?
> Worker: Motivation.
> Trainer: What leads to motivation?
> Worker: Reward.
> Trainer: No, rewards only work for a little while.
> *(Yasukawa et al. 2014, 400)*

The trainer then refocused the workers ways of measuring the cost of the problems, and therefore how much benefit could be gained by eliminating or minimising these problems.

Learning affordances for critical numeracy practices in workplaces

The two case studies, casual academics in the university sector, and production workers in lean production firms, provide examples of contrasting knowledge-producing activity systems. In the first example, the workers were questioning the contradictions between their existing work process and what they were gaining for themselves for the very practical motive of earning a living. In the second example, the workers were developing or further refining new work process knowledge, but not being supported in exploring the contradictions between the resultant productivity increases for the company and their static working conditions and pay.

Both activity systems produced new numeracy practices. The casual academics learnt that their rate of pay was determined by a model of work that involved far fewer hours of preparation, student consultations and marking than the actual hours that they worked and in fact were needed to conduct their work properly. They learnt to keep records of their time in ways that enabled them to compare their hours with each other, and importantly, to use them as evidence that they were not being fully paid for the work they did. However, these numeracy practices of recording their worked hours and making numerical comparisons with the model of pay in order to mount a dispute involved more than learning numeracy skills. The casual workers learnt that they needed to exercise activism as union members, and that by getting the union to lodge a pay dispute on their collective behalf rather than negotiating with their supervisors as individual casual employees, the dispute was both more powerful and safer for the casual tutors. While casual academics' struggles for industrial justice are far from over even ten years after this case study, this dispute and similar ones nationally helped the union and its casual academic members to win incremental improvements in their working conditions; for example, the marking of assessment tasks undertaken by casual academics are now paid as a completely separate item on the contract for the delivery of tutorials.

For the production workers learning the numeracy to implement lean production processes, the learning also involved recording the work they did. They plotted the number of finished products they produced each day, week and month, and graphically displayed them on a bar graph on a daily basis to compare against their target. But the motive for this activity system was to increase productivity levels for the company, that is, to implement new work process knowledge. There was little affordance for learning and producing more powerful negotiating knowledge for their own benefit as employees. While encouraging the workers to come up with suggestions for improving their work practices by producing more while reducing waste (of time and material resources), the trainers in each case ensured that the improvements were privileging productivity increases. Suggestions, such as that which emerged at Insulation Products which involved the company incurring a cost to achieve an improvement, were quickly silenced.

From a CHAT perspective, the differences in the objects or motives for the two knowledge-producing activity systems explain a large part of why learning unfolded differently in the three work sites. However, attention to other elements of the respective activity systems are useful in understanding more fully what is needed for workplaces to be sites of learning for critical numeracy practices.

CHAT places a focus on the tools and instruments that mediates the activity. In some instances the tools and instruments may be material artefacts (for example, a calculator, a ruler) and in some other instances they may be symbolic (for example, a formula, a set of instructions). In all three activity systems (two in the lean production firms, one in the university), there was, in addition to calculators, timekeeping devices, spreadsheets and other symbolic and material artefacts, a human mediator – the union delegate in the first case study, and lean production trainers engaged by the employers in the second case studies. In each of the three activity systems, there was a mediator who built on knowledge elicited from the workers; without this, the union dispute could not be substantiated, and improvements that would yield greater productivity in the workplaces could not be identified. In that sense, the pedagogy reflected respect for and acknowledgement of the workers' knowledge and experience; none of these learning sessions were facilitated by what could be characterised as a didactic teacher-centred pedagogy. There was in all cases the building of a community so that workers experienced what Engeström (2001) called the collective learning journey through the ZPD.

In all cases, the workers' existing activity systems had been disturbed: for the casual academics, their visibility in their union had been raised as a result of a shift in the union's prioritisation of casual academic members' concerns. Against the backdrop of increasing workload and continuing precariousness of their employment, the casual academics saw this disturbance as an opportunity to challenge their university's method of paying for their labour. It generated a new negotiating knowledge-producing activity system, mediated by the union delegate, to redress industrial injustice. As noted earlier, the dispute that was mounted and won, and similar activism in other institutions led to a radical transformation of the method of paying casual academics.

In the manufacturing firms, the disturbance to the workers' activity system was the introduction of lean production methods. This disturbance had already led to expansive learning in Hearing Solutions, to the extent that new work processes were visibly embraced, for example in the form of team business boards, performance targets and daily team meetings. The research did not extend long enough to determine whether workers in Insulation Products eventually embraced these new work practices or whether Hearing Solutions workers continued to fine-tune their work processes to achieve greater efficiencies. What was evident in the research was that collective learning in these companies was focused on generating and establishing new work process knowledge for purposes identified by their respective employer.

In neither of the case studies were the human mediators supporting the workers in an impartial way. The second lean production study points to the kinds of conflict identified by Gallo (2004) that workplace educators face if they want to show solidarity with the workers while being paid by the employer to meet their needs. If either of the lean production trainers felt conflicted in their roles, they did not make this evident, and in training at Insulation Products, the trainer actively steered the workers from discussing any solutions motivated by improvements in working conditions that were in possible conflict with the employer's agenda. The union activist, on the other hand, was engaged by the workers to support them in their cause, and was not bound to any allegiance to the workers' employer. Critical discussions were both possible and necessary to achieve the object of the casual academics' learning.

Conclusion

The chapter examined the possibilities and limitations of workplaces as sites of learning critical numeracy practices: numeracy practices that enable workers to understand, question or challenge the power relations in their workplace. The case of the casual academics demonstrated that workplaces can be sites of critical numeracy practices, while the case of the lean production training demonstrated that workplaces can also suppress the development of critical numeracy practices. The distinction offered by Worthen (2008) between activity systems that produce negotiating knowledge and those that produce (only) work process knowledge helps us to see the motive of the activity systems that lead to critical insights about workers' interest in earning a living on the one hand, and knowledge that enables the employer to get more out of their workers' labour on the other.

In many of the human capital discourses, claims are made about numeracy and literacy being 'key information processing skills' for workers in the twenty-first century (OECD 2013, 3), and these claims are turned around to attribute poor literacy or numeracy as reasons when adults struggle to secure work or remain in low-paid positions. While some governments, including the Australian government, fund literacy and numeracy programmes to improve the employability of the long-term unemployed, it is questionable how many of such programs treat numeracy as a critical resource for developing negotiating knowledge for the workplace. In the author's observations of numerous numeracy and literacy classes in Australian labour market programmes, there have not been any that was teaching learners to examine the power relations they would experience in the workplace in critical ways. Jacobson's (2016) call for radical education seems currently elusive in these labour market programmes, but for that reason education for critical numeracy is urgent, particularly for learners likely to be working in one of the growing numbers of non-unionised workplaces. Engeström's (2001) CHAT framework provides resources for analysing the factors that can lead to affordances for critical numeracy learning and the development of critical workplace practices; this analysis in turn, could perhaps also inform and extend what Evans et al. characterised as a numerate environment into a critically numerate environment.

References

Baker, D. (1998) "Numeracy as social practice" *Literacy and Numeracy Studies*, 8(1): 37–50.
Belfiore, M.E., Defoe, T.A., Folinsbee, S., Hunter, J. and Jackson, N.S. (2004) *Reading work: Literacies in the New Workplace* London: Routledge.
Black, S., Yasukawa, K. and Brown, T. (2013) *Investigating the 'Crisis': Production Workers' Literacy and Numeracy Practices* Adelaide, SA: National Centre for Vocational Education Research,.
Bond, M. (1999) "What about the men? Reflections from a picket line on returning to learning" *Studies in the Education of Adults*, 31(2): 164–180
Bond, M. (2000) "Understanding the benefits/wages connection: financial literacy for citizenship in a risk society" Studies in the Education of Adults, 32(1): 63–77.
Brown, T., Goodman, J. and Yasukawa, K. (2006) *Getting the Best of You for Nothing: Casual Voices in the Australian Academy* Melbourne: National Tertiary Education Union

(NTEU). Retrieved from https://www.nteu.org.au/library/view/id/422 (accessed 19 January 2018).

Brown, T., Goodman, J. and Yasukawa, K. (2010) "Academic casualization in Australia: Class divisions in the university" *Journal of Industrial Relations*, 52(2): 169–82.

Engeström, Y. (2001) "Expansive learning at work: Toward an activity theoretical reconceptualization" *Journal of Education and Work*, 14(1): 133–56.

European Union High Level Group of Experts on Literacy (2012) *Final report: September 2012*. Retrieved from http://ec.europa.eu/dgs/education_culture/repository/education/policy/school/doc/literacy-report_en.pdf (accessed 25 May 2017).

Evans, J. (1999) "Building bridges: Reflections on the problem of transfer of learning in mathematics" *Educational Studies in Mathematics*, 39(1): 23–44.

Evans, J., Yasukawa, K., Mallows, D. and Creese, B. (2017) "Numeracy skills and the numerate environment: Affordances, opportunities, supports and demands". *Adults Learning Mathematics: An International Journal*, 12(1), 17–26

FitzSimons, G. E. (2005) "Numeracy and Australian workplaces: Findings and implications" *Australian Senior Mathematics Journal*, 19(2): 27–40.

Gallo, M.L. (2004) *Reading the World of Work: A Learner-Centered Approach to Workplace Literacy and ESL* Malabar, FL: Krieger Publishing.

Harris, M. (ed.) (1991) *Schools, Mathematics and Work* London: The Falmer Press.

Hoyles, C., Noss, R., Kent, P. and Bakker, A. (2010) *Improving Mathematics: The Need for Techno-Mathematical Literacies* London: Routledge.

Hull, G.(ed.) (1997) *Changing Work, Changing Workers: Critical Perspectives on Language, Literacy, and Skills* Albany, NY: SUNY Press.

Jacobson, E. (2016) "Workforce development rhetoric and the realities of 21st century capitalism" *Literacy and Numeracy Studies*, 24(1): 3–22.

Kane, P.J. (2014) "An investigation into estimation and spatial sense as aspects of workplace numeracy: a case study of recycling and refuse operators within a situated learning model" MPhil dissertation, Auckland University of Technology.

Kanes, C. (2002) "Towards numeracy as a cultural historical activity system". In Valero, P. and Skovsmose, O. (eds), *Proceedings of the Third International Mathematics Education and Society Conference* Helsingør, Denmark, 2–7 April. Copenhagen: Centre for Research in Learning Mathematics.

Lave, J. (1988) *Cognition in Practice: Mind, Mathematics and Culture in Everyday Life* Cambridge: Cambridge University Press.

Nunes, T., Schliemann, A.D. and Carraher, D.W. (1993) *Street Mathematics and School Mathematics*. Cambridge: Cambridge University Press.

OECD (2013) *OECD Skills Outlook 2013: First Results from the Survey of Adult Skills* Paris: OECD Publishing.

Wedege, T. (2000) "Mathematics knowledge as a vocational qualification". In Bessot, A. and Ridgway, J. (eds), *Education for Mathematics in the Workplace* Dordrecht: Springer.

Wedege, T. (2002) "'Mathematics – that's what I can't do': People's affective and social relationship with mathematics" *Literacy and Numeracy Studies*, 11(2): 63–78.

Williams, J. and Wake, G. (2007) "Black boxes in workplace mathematics" *Educational Studies in Mathematics*, 64(3): 317–343.

Worthen, H. (2008) "Using Activity Theory to understand how people learn to negotiate the conditions of work" *Mind, Culture, and Activity*, 15(4): 322–38.

Yasukawa, K. and Black, S. (2016) "Policy making at a distance". In Yasukawa, K. and Black, S. (eds) *Beyond Economic Interests: Critical Perspectives on Adult Literacy and Numeracy in a Globalised World* Rotterdam: SensePublishers.

Yasukawa, K. and Brown, T. (2012) "Bringing critical mathematics to work: But can numbers mobilise?". In Skovsmose, O. and Greer, B. (eds), *Opening the Cage: Critique and Politics of Mathematics Education* Rotterdam: SensePublishers.

Yasukawa, K., Brown, T. and Black, S. (2013) "Production workers' literacy and numeracy practices: Using cultural-historical activity theory (CHAT) as an analytical tool" *Journal of Vocational Education & Training*, 65(3): 369–84.

Yasukawa, K., Brown, T. and Black, S. (2014) "Disturbing practices: Training workers to be lean" *Journal of Workplace Learning*, 26(6/7): 392–405.

Zevenbergen, R.J. (2011) "Young workers and their dispositions towards mathematics: Tensions of a mathematical habitus in the retail industry" *Educational Studies in Mathematics*, 76(1): 87–100.

Conclusion

14

EXPANDING AND DEEPENING THE TERRAIN

Numeracy as social practice

Kara Jackson, Alan Rogers and Keiko Yasukawa

In this volume, we have seen how numeracy learning is lifelong and lifewide. The volume has featured research not only in school contexts but in post-compulsory education as well. But it is not only educational settings that are the site of numeracy practice research. We have seen accounts of numeracy practices in workplaces, in families and in a social movement. In Chapter 1, the authors mapped the 'terrain' of social practices perspectives of numeracy, identifying and elucidating the contributions of key theoretical influences that have informed research in numeracy as a social practice. In this final chapter, the editors revisit the terrain in light of the salient themes in the contributions made in the chapters in this volume. The editors first discuss some of the emergent themes and then draw out implications for policy and practice.

The 'invisibility' of mathematics in everyday practices

Much research that reflects a numeracy as social practice perspective is motivated, in part, to make visible the mathematics that 'everyday' people do, in their everyday lives – and thus to position people as competent and capable of engaging in mathematics. This presents a challenge to the all too frequent deficit (or, worse, crisis) discourse that surrounds numeracy and literacy in many countries, when league tables are published on international tests such as the Programme for International Student Assessment (PISA) for school children and the Programme for the International Assessment of Adult Competencies (PIAAC) for adults. In these tests, the mathematics that people *do* outside the walls of the formal mathematics classroom is invisible or insignificant. People who, from the gaze of researchers taking a numeracy as social practice perspective, are engaging in out-of-school numeracy practices, often also suggest they are not 'really doing mathematics' and underestimate their mathematical knowledge and skills (Wedege 2002).

The authors in this volume illustrate how everyday numeracy practices are rendered invisible. For example, the invisibility and active invisibilisation of everyday numeracy practices in school education is highlighted in the chapters by Barwell and by Alshwaikh and Yasukawa. In both cases the framing of the content and the pedagogical assumptions of the textbooks used in Canada and the occupied Palestine, respectively, left little room for learners' voices and numeracy practices to enter the classroom.

Discontinuities and invisibilisation are also seen in other contexts. Both Alangui and Kane identify mathematical skills and concepts that are present in work practices of building stone walls, and management of orchards and refuse/recycling operations, respectively. In these work practices, skilled estimation of a range of quantities and measures is needed; yet workers either did not notice their work as being mathematical or discounted the degree of mathematical skills and thinking their work involved.

Similarly, Yasukawa's chapter on numeracy practices in workplaces points to active invisibilisation of the connections between productivity and pay. In the case of the university workers, the invisibilisation was challenged by the workers, while in the manufacturing companies, the workers were prevented from contemplating the connections through the pedagogical manoeuvres of the trainers.

Numeracy practices are fluid, unstable and context-contingent

The discussions in the chapters in this volume also indicate that 'everyday' numeracy practices are not standardised or static. Instead, they are fluid, unstable and context-dependent.

Many factors create the changing nature of everyday numeracy practices. First, of course, is that each individual constantly enters new situations calling for new learning during the course of their lives – for example, as worker, as family member, as community member, as member of a religious group. What is more, the interpretation of such roles change – a mother or father in one generation may be very unlike a mother or father in a previous generation. Like the wage-earners in Mexico (Kalman and Solares), they will face different financial situations, calling for new relationships with others which will mediate literacy and numeracy practices. All of these changing conditions call for new learning – and new numeracy practices.

Second, the cultural context changes –sometimes very rapidly. Alshwaikh points out how political changes affect the ways in which numeracy is learned. One major change in recent years has been the spread of new information and communication technologies, even into remote parts of Africa and Asia. For example, throughout Africa, especially in areas away from urban centres, people are using mobile phones to bank money – to make deposits and withdraw as and when they need it – calling for sophisticated numeracy practices. Other socio-economic-cultural-environmental changes demand new numeracy practices. Although the stone wall builders in the Philippines may have changed their traditional methods of construction slowly, they did change. The kiwifruit pickers faced changing climatic conditions that affected yields.

Third, the natural innovativeness of people means that they will construct for themselves new ways of acting out numeracy, new ways of solving problems, short cuts. The waste collection drivers in Kane's chapter were constantly making new decisions based on numeracy practices. The stone wall builders in Alangui's chapter adapt and innovate on practices that have developed over centuries. Such innovations build on prior learning and are often learned without any special constructed learning programme; they are learned partly by trial and error, partly by asking and copying others, sharing insights and practices, using the natural scaffolding of the community. In other words, people build up tacit funds of knowledge to negotiate life; and these funds of knowledge lay foundations for new planned and purposeful learning (Rogers 2014a; Robinson-Pant 2016; see also Eraut 2000; Livingstone 2001; Straka 2004).

Numeracy practices are local and global

While a social practice perspective on numeracy uncovers the rich, situative nature of numeracy practices, these practices are not produced and retained within their contexts. There is resonance with the analysis of literacy practices in the paper 'Limits of the local' (Brandt and Clinton 2002). As several chapters show, some numeracy practices – especially those associated with formal schooling – are remarkably stable across time and space.

Authors describing compulsory mathematics education in Palestine (Alshwaikh and Yasukawa), apartheid South Africa (Khuzwayo), and primary education in India (Rampal) and Nepal (Shiohata) all describe pedagogy and curricula that privilege students memorising sets of procedures to solve routine problems. The teacher and text are positioned as having mathematical authority, and the students are positioned as recipients of knowledge.

Moreover, authors in this volume describe how persistent and cemented compulsory mathematics schooling practices are. For example, Khuzwayo showed how the enduring influence of mathematics education in the apartheid era is still being felt over two decades after the collapse of the apartheid system in South Africa, taking the form of an 'occupation of minds'. While the curricula have seen changes, they have been changed from the top by 'experts', continuing to position teachers as ignorant players. Similarly, Rampal described revolutionary changes to school mathematics textbooks in India in 1990, in which textbook developers made everyday situations that were grounded in the lives of real Indian citizens the focus of mathematisation. The textbook developers took deliberate measures to broaden who was positioned as capable of engaging in mathematics in the story problems. However, Rampal also details the difficulty in supporting teachers to develop the pedagogy that is necessary to employ the texts in a way that broadens conceptions of both what counts as mathematics and who can do mathematics. Shiohata underscores and amplifies the challenges Khuzwayo and Rampal point to. She describes efforts to interrupt business-as-usual primary teaching practices in Nepal through a focus on formative assessment, aimed at uncovering students' ideas and building instruction in response to those ideas. However, as she details, in large part, the efforts had little traction. While disappointing, these findings are not surprising. Conventional school mathematics practices have developed and accreted over centuries; nudging them (let alone displacing them) in any substantial way requires sustained, coordinated action regarding structures, pedagogy and resources.

Both humans and objects participate in numeracy practices

As we highlighted in Chapter 1, tools shape numeracy activity – all of the chapters illustrated this point. In fact, what we see in this volume reflects Latour's (1987) observation regarding how both humans and objects direct activity. Attempting to overcome a subject/object dualism, Latour argues that material resources can be conceptualised as *actants*. Just as a human actor can play multiple roles or act as a particular kind of person in one situation but differently in another, an actant can serve different goals in different situations. Latour's point is not that a material object has *all* the same properties as humans, but that action is not the property of one agent (material or human). Rather, 'action is a property of associated entities. … Action is simply not a property of humans but of an association of actants' (Latour 1987, 35).

In this volume, one object that was seen to shape pedagogical practices in mathematical learning was the textbook, in Palestinian (Ashwaikh and Yasukawa) and Indian classrooms (Rampal). We saw too in Ashwaikh and Yasukawa's, and in Barwell's chapter, how language is a powerful actant in numeracy learning contexts. In fact in both cases, the language of the classroom was shown to shape classroom practices that worked to exclude learners' out of school numeracy practices.

As noted by Street and Baker (2006), numeracy practices are multimodal, relying on people's use of oral, written and symbolic language (particularly in school mathematics), material artefacts such as measuring devices (Boistrup, Bellander and Blaesid; Kane) and other equipment such as the hydraulic arm for lifting garbage bins (Kane). In his study of workers in the USA, Rose (2004) talked about carpenters' use of 'material mathematics' and the use of multiple senses of the ear, the eye and the body as a whole: for example, to hear the sound of the power tools to check they are working properly; to judge that structures reflect the type of symmetry that is desirable; and to ensure that the right amount of pressure is put on the saw when cutting wood. The kind of 'disciplined perception' (Rose 2004, 73; see also Stevens and Hall 1998) that Rose notices among the carpenters is what has been illustrated in studies of workplace numeracy in this volume. They are also multisensory and require this disciplined perception based on experience and a deep understanding of the purpose and context of the numeracy practices. The studies in this volume suggest that the further away from the school classroom, the more multimodal and multisensory the numeracy practices are. This can also contribute to the invisibilisation of mathematics in numeracy practices.

As the studies of the use of tools in this volume have also shown, there are traditions surrounding the use of tools – whether they are new digital technologies or more manual tools – that shape the numeracy practices within particular social practices. As Strässer (2007) and others have argued, new technologies have led to many mathematical functions to be 'black-boxed', that is, out of the direct control of the people operating the technologies (for example, electronic kitchen scales, formulae built into spreadsheets, the calculator function of cell phones). This does not necessarily mean that the operators are lacking in numeracy skills. Rather, they are participating in new forms of numeracy practices, working with these technologies to accomplish tasks that were historically accomplished differently.

Numeracy practices are political

An assumption of a numeracy as social practice perspective is that power relations inevitably shape the uses of mathematics in practice, the meanings that people ascribe to their numeracy practices as well as how 'outsiders' ascribe meanings to their practices. The chapters in this volume illustrate this point – and enrich the field's understanding of how power operates in a variety of contexts and in different ways.

Khuzwayo, in his study of compulsory mathematics education in apartheid South Africa, highlights the racialised nature of numeracy practices. Through historical analysis, he shows how mathematics education was explicitly used as a tool to oppress Blacks – including how Blacks were explicitly constructed as unable to engage in mathematics.

In a study of casual academics in an Australian university, Yasukawa sheds light on how numeracy practices can be used to uncover and act against oppressive power relations in the workplace, when paired with 'negotiating knowledge'. However, she also shows how in the absence of negotiating knowledge, numeracy practices are not enough to challenge unfair labour practices.

The chapter by Knijnik and Wanderer highlights the politics of knowledge and the question of whose mathematics counts and whose doesn't. They illustrate how Western school mathematics assumes the status of universal legitimacy because of its pursuit of transcendence, untainted by the particularities of the real-life contexts of people, while the ethnomathematics is deemed inferior owing to its contingency on the specific, local context in which it is used. For the Brazlian landless people, it is the latter numeracy practices that have emerged in response to the kinds of situations and problems they encounter. This makes it more readily powerful, in terms of solving problems impinging on these people's livelihoods, compared with the forms of practice that are valued in formal school curricula.

Implications for numeracy policy and pedagogy

The implications of this volume for adult numeracy and mathematics educators are profound. As illustrated by Alshwaikh and Yasukawa, Barwell, Khuzwayo, Knijnik and Wanderer, Rampal, and Shiohata, what has been described in the Introduction as the autonomous model of numeracy permeates formal schooling in mathematics around the world. The teacher designs lessons for the content or body of knowledge to be taught. The content is specified in the curriculum, which is seen as fixed and is devised by the education system in that society. Learners are introduced to the concepts involved, they then practise these skills and finally they use the skills to tackle problems chosen by the teacher. In this model of teaching and learning, the teacher is the holder of the knowledge and the learner is the receiver, and the learner performs activities that may not be meaningful to them.

Moreover, at a time when national education policies are increasingly influenced by transnational initiatives, in particular, large-scale testing such as PISA and PIAAC, there is a constant risk of numeracy being reduced to the autonomous conception of transferable cognitive skills. This volume has documented numerous examples that illustrate that everyday numeracy practices can have little resemblance to what we might find in these international tests. These 'stories from the field' can be used to nuance the stories told by the test statisticians – to tell stories that may shape policies that inform pedagogies in socially and culturally sustaining ways.

Recognising the limitations of an autonomous model of mathematics teaching and learning, what do the chapters suggest for the reform of pedagogy? Broadly speaking, all of the chapters support the premise that an approach to numeracy/mathematics education that is built on a deficit model misses the mark. People are already users of multiple numeracy practices. It is, of course, possible to introduce them to new ways of doing numeracy – in much the same way as a neighbour would show how to interpret the words and numbers on a mobile phone card, not as a decontextualised exercise but out of a real and immediate sense of need to top-up the balance on one's phone.

Prior studies of people engaging in everyday numeracy practices have resulted in researchers suggesting the value in integrating such everyday practices or 'funds of knowledge' into schooling (compulsory and adult education) contexts (see for example, Baker and Rhodes 2007; Civil 2002). Similarly, in this volume, Alangui calls for everyday numeracy practices, like those of the stone wallers, to be made more visible in formal mathematics learning contexts because it would both enrich and make mathematics learning in formal settings more relevant for the students. Rampal describes a substantial effort to design and implement new primary mathematics textbooks in India that reflected everyday practices. And Lekoko and colleagues describe a pedagogical model being used in adult numeracy education in Botswana that enables learners to 'see' mathematics embedded in their daily practices. Learners are then supported to learn formal mathematical language, symbols and so forth to characterise what they already do. The work that Rampal and Lekoko and colleagues. describe is laudable – they are engaging in sustained, strategic efforts to disrupt business as usual.

On the other hand, Kane's chapter problematises the call to make workplace practices visible in formal mathematics schooling curricula. Because the workers' numeracy practices are embedded in workplace practices, the value and feasibility of meaningfully extracting aspects to incorporate into a formal mathematics curriculum are questionable. The notion of extracting mathematical aspects from a practice suggests that different numeracy practices are constituted by some straightforward insertion of mathematical elements into an 'incompletely' constituted practice, a reversion to the 'autonomous model' of numeracy that sits in tension with the notion of numeracy as social practice (Baker 1998). Boistrup and colleagues' chapter also raise questions about the value of extracting aspects of workplace practices to incorporate into formal mathematics curricula. In detailing the relations between 'school mathematics' and construction work, Boistrup and colleagues illustrate that construction workers recontextualise school mathematics in the context of construction activity (e.g., using procedures learned in school mathematics to calculate the number of lorries necessary to remove sand from a site). However, they also illustrate that aspects of a focal construction activity (like removal of sand) are specific to construction (e.g., mathematically modelling the mass of sand to be removed depends on the characteristics of the ground). In other words, while there is a place for integrating school mathematics and workplace mathematics in vocational education, simply applying school mathematics to construction contexts is inadequate for supporting construction workers' development of workplace numeracy practices.

It is crucial, however, whether it is formal school curricula or workplace training, that learners are viewed and treated as having numeracy resources on which to build. As one example, in the adult and vocational education sector in Australia, New Zealand and the UK, there has been much concern about the numeracy (and literacy) levels of workers and students in the vocational education sector. This has led to initiatives to 'embed' or 'integrate' numeracy, literacy and (English) language learning into the vocational courses. Researchers working with literacy as social practice (LSP) perspective suggest that embedded numeracy, literacy and language is an approach that concurrently teaches these skills in the vocational programs and 'provides learners with the confidence, competence and motivation necessary for them to succeed in qualifications, in life, and at work' (Roberts et al. 2005, 5). Similarly, Ivanic et al. (2009) facilitated an action research project with vocational teachers in further education colleges in the UK that shifted the ways in which literacy (mostly) and numeracy were conceptualised by teachers. The teachers had, on the one hand, a highly deficit view of their learners' literacy and numeracy levels, while on the other hand, they also had a very narrow, academic view of what constituted the literacies and numeracies that were going to be relevant to the students when they enter the workforce in their chosen fields. Learners in this research kept a log of their everyday literacy and numeracy practices which, when they presented, revealed a much larger repertoire of literacy practices than the teachers had assumed, much to their teachers' surprise. This was a catalyst that led some of the teachers to question their own assumptions about the kinds of literacies and numeracies they needed to be building on and teaching the students, so that these reflected the different contexts of literacy and numeracy that the students were negotiating: for negotiating the administrative requirements of being a college student, for learning the subject content, for assessment, for becoming a member of the workforce in their chosen field. Their 'Literacies for Learning in Further Education' (n.d.) framework was an outcome of the project that teachers could use to think through the literacy and numeracy that would be privileged and foregrounded in the different contexts of the students' college life.

In the adult and vocational education sector in developing countries, there has been less concern with numeracy on its own; it has usually been seen as a subset of the adult literacy curriculum. And this is despite the many efforts made to 'embed' literacy learning within the development of skills (especially livelihood and citizenship skills) in the many different forms of 'functional literacy' and adult literacy with vocational skills at grass-roots levels (Rogers 2014b). And yet in many of these programmes, it is the numeracy practices that many students demand and which come to feature substantially.

One example is the LETTER (Learning for Empowerment Through Training in Ethnographic Research) programme developed by Alan Rogers and Brian V. Street at the request of Nirantar, an NGO in India committed to women's education for women's empowerment in 2000 (Nirantar 2007). The aim was to develop a practical training programme for adult literacy teachers to use ethnographic approaches to learn about the existing everyday literacy practices of the literacy learners and their communities, so that these could be used as the basis for teaching other (more formal) literacy practices. There was no specific numeracy component – numeracy was seen as a component of the literacy curriculum. But the workshops in Delhi in 2006, facilitated by Brian V. Street and Dave Baker, given Dave Baker's experiences and interests (e.g. Baker 1998), focused on the existing numeracy practices of the literacy learners. Community members were found to use multiple and locally determined ways of measuring and calculating that were different from those of their teachers and those of the adult literacy curriculum, and yet were so commonplace that no one had even remarked on them. These became the subject of localised small-scale research projects which all the trainees undertook as part of the training.

The programme was taken to Ethiopia (see Gebre et al. 2009) and Uganda (see Openjuru et al. 2016). In both, the small-scale ethnographic-style research projects into everyday literacy practices found many local ways of measuring weights, sizes, time and distances and local ways of calculating, using language which was very local and not to be found in the adult literacy textbooks (see also Saraswathi 2012). One finding across the studies was that when local practices were examined ethnographically, those who used them drew no distinction between literacy and numeracy practices, as for example with the uses of mobile phones which combine letter and number literacy practices. It seems to be academics and educators who draw a distinction between literacy and numeracy practices; real life is more integrated.

Across educational programmes, those who are in relative positions of power determine which and whose knowledge is valuable; and often, local knowledge is discounted. To help confront these political realities, Degener (2001) argues that adult literacy and numeracy programmes should not only offer adult learners opportunities to learn specific literacy skills and practices, but there should also be 'contextualised instruction within a framework of social activism and societal transformation' (Degener 2001, 26). For numeracy in particular, Yasukawa and Johnston adopt a more expansive stance suggesting that 'it cannot be called numeracy unless it is political, concerned with issues of power' (Yasukawa and Johnston 1994, 198).

This is true for teacher education programs as well as adult numeracy programs. Turner and colleagues (2014) have created modules for pre-service mathematics methods courses in which prospective primary teachers in the US are supported to investigate their students' numeracy practices, as well as other cultural and linguistic practices in an effort to deliberately design instruction that builds on their students' multiple knowledge bases.

Rochelle Gutiérrez (2012; 2013) argues that mathematics teachers in compulsory education need to develop 'political knowledge' in addition to pedagogical knowledge, content knowledge, and knowledge about students' communities and cultural practices. Political knowledge entails:

> negotiating the world of high stakes testing and standardisation, connecting with and explaining mathematics to community members and district officials, and buffering oneself, reinventing, or subverting the system in order to be an advocate for one's students.
>
> *(Gutiérrez 2012, 33)*

This resonates with Worthen's (2008) notion of 'negotiating knowledge' that Yasukawa referred to in her chapter that explored the possibilities of workers developing critical numeracy practices to make sense of the contradictions between their employer's demands for increased productivity and their unchanging poor wages. Teachers, like workers described by Yasukawa, are caught between the need to work within the compulsory education system and the need to work in a way that makes sense to them – philosophically, pedagogically and politically.

In closing, the contributions that this volume makes to support teachers, facilitators of learning and curriculum writers across learning contexts include the following. First, this volume has highlighted ways of researching and shining light on learners' numeracy practices that can be used to help learners make more connected meanings with the mathematics they have to learn in various systems and programmes. Second, this volume has provided research evidence that challenges the autonomous model of numeracy and the underlying assumption that mathematics taught in school can be transferred into the varied numeracy practices required in out of school contexts. Third, this volume provides resources that can help develop pre-service and current teachers' 'negotiating knowledge'. And fourth, the volume provides resources that can help teachers to see, listen to, engage with and build on their learners' everyday numeracy practices, thereby producing a richer picture of numeracy learning than the current forms of high-stakes testing data can provide. This, we hope, will go some way towards challenging existing policies.

References

Baker, D. (1998) "Numeracy as social practice" *Literacy and Numeracy Studies*, 8(1): 37–50.

Baker, D. and Rhodes, V. (2007) *Making Use of Learners' Funds of Knowledge for Mathematics and Numeracy: Improving Teaching and Learning of Mathematics and Numeracy in Adult Education* NCETM/Maths4Life. Retrieved from www.ncetm.org.uk/files/254456/researchfundsofknowledge.pdf (accessed 9 July 2008).

Brandt, D. and Clinton, K. (2002) "'Limits of the local': expanding perspectives on literacy as social practice" *Journal of Literacy Research*, 34(3): 337–356.

Civil, M., (2002) "Chapter 4: Everyday mathematics, mathematicians' mathematics, and school mathematics: can we bring them together?" *Journal for Research in Mathematics Education*, 11: 40–62.

Degener, S. C. (2001) "Making sense of critical pedagogy in adult literacy education", *Review of Adult Learning and Literacy: Volume 2*, Boston, MA: NCSALL.

Eraut, M. (2000) "Non-formal learning, implicit learning and tacit knowledge in professional work". In Coffield, F. (ed), *The Necessity of Informal Learning* Bristol: Policy Press.

Gebre A., Rogers, A. and Street, B. V. (eds) (2009) *Everyday Literacies in Africa: Ethnographic Studies of Literacy and Numeracy in Ethiopia* Retrieved from http://www.bald.org.uk/wp-content/uploads/2012/12/Ethiopia-whole-book-knj.pdf (accessed 18 January 2018).

Gutiérrez, R. (2012) "Embracing Nepantla: Rethinking 'knowledge' and its use in mathematics teaching" *REDIMAT – Journal of Research in Mathematics Education*, 1(1): 29–56.

Gutiérrez, R. (2013) "Why (urban) mathematics teachers need political knowledge" *Journal of Urban Mathematics Education*, 6 (2): 7–19.

Ivanic, R., Edwards, R., Barton, D., Martin-Jones, M., Fowler, Z., Hughes, B., Mannion, G., Miller, K., Satchwell, C. and Smith, J. (2009) *Improving Learning in College: Rethinking Literacies across the Curriculum* London: Routledge.

Latour, B. (1987) *Science in Action: How to Follow Scientists and Engineers through Society* Cambridge, MA: Harvard University Press.

Literacies for Learning in Further Education (n.d.). "Curriculum review using the LfLFE framework" Retrieved from http://www.lancaster.ac.uk/lflfe/publications/pubsdocs/LfLFE%20Framework%20from%20booklet.pdf (accessed 11 September 2017).

Livingstone, D. W. (2001) "Adults' informal learning: definitions, findings, gaps and future research" https://tspace.library.utoronto.ca/handle/1807/2735 (accessed 12 January 2006).

Nirantar (2007) *Exploring the Everyday: Ethnographic Approaches to Literacy and Numeracy* New Delhi: Nirantar and ASPBAE. Retrieved from http://www.nirantar.net/uploads/files/Part1.pdf; http://www.nirantar.net/uploads/files/part2.pdf and http://www.nirantar.net/uploads/files/part3.pdf (accessed 18 January 2018).

Openjuru, G., Baker, D., Rogers, A. and Street, B. V. (2016) *Exploring Adult Literacy and Numeracy Practices: Ethnographic Case Studies from Uganda* Bury St Edmunds: Uppingham Press. Retrieved from http://www.uppinghamseminars.co.uk/Exploring%20Adult%20Literacy.pdf (accessed 18 January 2018).

Roberts, C., Gidley, N., Eldred, J., Brittan, J., Grief, S., Cooper, B., Gidley, N., Windsor, V., Eldre, J., Castillino, C. and Walsh, M. (2005) *Embedded Teaching and Learning of Adult Literacy, Numeracy and ESOL: Seven Case Studies* London: National Research and Development Centre for Adult Literacy and Numeracy.

Robinson-Pant, A. (2016) *Learning Knowledge and Skills for Agriculture to Improve Rural Livelihoods* Paris: UNESCO. Retrieved from http://unesdoc.unesco.org/images/0024/002457/245765e.pdf (accessed 9 Novermber 2017).

Rogers, A. (2014a) *The Base of the Iceberg: Informal Learning and its Impact on Formal and Non-Formal Learning,* Study Guides to Adult Education. Leverkusen: Barbara Budrich.

Rogers A. (2014b) *Skills Development and Literacy: Ethnographic Challenges to Policy and Practice,* Working Paper 2, Norwich: Centre for Applied Research in Education (CARE), University of East Anglia. Retrieved from https://www.uea.ac.uk/documents/595200/0/CARE+Working+Paper+2+Rogers.pdf (accessed 18 January 2018).

Rose, M. (2004) *The Mind at Work: Valuing the Intelligence of the American Worker* New York: Penguin.

Saraswathi, L. S. (2012) *Everyday Mathematics and the Classroom: Case Studies from Rural South India.* Retrieved from http://www.balid.org.uk/wp-content/uploads/2012/10/Saraswathi-Everyday-mathematics-and-the-classroom.pdf (accessed 18 January 2018).

Strässer, R. (2007) "Everyday instruments: On the use of mathematics". In Blum, W., Galbraith, P. L., Henn, H. L. and Niss, M. (eds), *Modelling and Applications in Mathematics Education* New York: Springer.

Stevens, R. and Hall, R. (1998) "Disciplined perception: Learning to see in technoscience". In Lampert, M. and Blunk, M. L. (eds), *Talking Mathematics in School: Studies of Teaching and Learning* Cambridge: Cambridge University Press.

Street, B. V. and Baker, D. (2006) "What about multimodal numeracies?". In Pahl, K. and Rowsell, J. (eds), *Travel Notes from the New Literacy Studies* Cleveden: Multilingual Matters.

Straka, G. A. (2004) "Informal learning: genealogy, concepts, antagonisms and questions". Retrieved from https://www.pedocs.de/volltexte/2014/9162/pdf/Straka_2004_Informal_learning.pdf (accessed 2 January 2006).

Turner, E. E., Aguirre, J. M., Bartell, T. G., Drake, C., Foote, M. Q. and Roth McDuffie, A. (2014) "Making meaningful connections with mathematics and the community: Lessons from pre-service teachers". In Bartell, T. G. and Flores, A. (eds), *Embracing Resources of Children, Families, Communities, and Cultures in Mathematics Learning* San Bernardino, CA: TODOS.

Wedege, T. (2002) "'Mathematics – that's what I can't do': People's affective and social relationship with mathematics" *Literacy and Numeracy Studies* 11(2): 63–78.

Worthen, H. (2008) "Using Activity Theory to understand how people learn to negotiate the conditions of work" *Mind, Culture, and Activity*, 15(4): 322–38.

Yasukawa, K. and Johnston, B. (1994) "A numeracy manifesto for engineers, primary teachers, historians … a civil society – can we call it theory?". In *Proceedings of the Australian Bridging Mathematics Network Conference* Sydney, NSW: Australian Bridging Mathematics Network.

INDEX

Notes are referenced as, for example, 72n3 – page 72, note 3

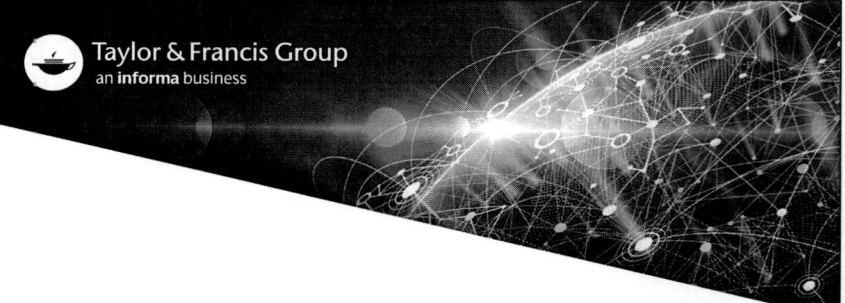